Supply Chain Design and Management

ACADEMIC PRESS SERIES IN ENGINEERING

SERIES EDITOR
J. DAVID IRWIN

Auburn University

This is a series that will include handbooks, textbooks, and professional reference books on cutting-edge areas of engineering. Also included in this series will be single-authored professional books on state-of-the-art techniques and methods in engineering. Its objective is to meet the needs of academic, industrial, and governmental engineers, as well as to provide instructional material for teaching at both the undergraduate and graduate level.

This series editor, J. David Irwin, is one of the best-known engineering educators in the world. Irwin has been chairman of the electrical engineering department at Auburn University for 27 years.

Published books in the series:
Power Electronics Handbook, 2001, M. H. Rashid, editor
Control of Induction Motors, 2001, A. Trzynadlowski
Embedded Microcontroller Interfacing for McoR Systems, 2000,
    G. J. Lipovski
Soft Computing & Intelligent Systems, 2000, N. K. Sinha, M. M. Gupta
Introduction to Microcontrollers, 1999, G. J. Lipovski
Industrial Controls and Manufacturing, 1999, E. Kamen
DSP Integrated Circuits, 1999, L. Wanhammar
Time Domain Electromagnetics, 1999, S. M. Rao
Single- and Multi-Chip Microcontroller Interfacing, 1999, G. J. Lipovski
Control in Robotics and Automation, 1999, B. K. Ghosh, N. Xi, and
    T. J. Tarn

SUPPLY CHAIN DESIGN AND MANAGEMENT

STRATEGIC AND TACTICAL PERSPECTIVES

MANISH GOVIL, PHD
i2 Technologies, Dallas, Texas

JEAN-MARIE PROTH, PHD
INRIA and University of Metz, France

ACADEMIC PRESS

An Elsevier Science Imprint

San Diego San Francisco New York Boston
London Sydney Tokyo

This book is printed on acid-free paper. ∞

Copyright © 2002 by ACADEMIC PRESS

All rights reserved.
No part of this publication may be reproduced or transmitted in any form or by any means, electronic or mechanical, including photocopy, recording, or any information storage and retrieval system, without permission in writing from the publisher.

Requests for permission to make copies of any part of the work should be mailed to: Permissions Department, Harcourt, Inc., 6277 Sea Harbor Drive, Orlando, Florida, 32887-6777.

Explicit permission from Academic Press is not required to reproduce a maximum of two figures or tables from an Academic Press chapter in another scientific or research publication provided that the material has not been credited to another source and that full credit to the Academic Press chapter is given.

Designations used by companies to distinguish their products are often claimed as trademarks or registered trademarks. In all instances in which Academic Press is aware of a claim, the product names appear in initial capital or all capital letters. Readers, however, should contact the appropriate companies for more complete information regarding trademarks and registration.

Academic Press
A Harcourt Science and Technology Company
525 B Street, Suite 1900, San Diego, California 92101-4495, USA
http://www.academicpress.com

Academic Press
Harcourt Place, 32 Jamestown Road, London, NW1 7BY, UK
http://www.academicpress.com

Library of Congress Catalog Card Number: 2001094751

International Standard Book Number: 0-12-294151-9

PRINTED IN THE UNITED STATES OF AMERICA
02 03 04 05 06 07 ML 9 8 7 6 5 4 3 2 1

To my wife, Fabienne, and my daughter, Stéphanie
  — Jean-Marie

To my grandmothers,
Shakuntala Devi and Gargi Devi
  — Manish

Contents

Foreword xi

1

Introduction 1

2

Definition of a Supply Chain 7

2.1. Global Definition   7
2.2. Examples and Counterexamples of Supply Chains   9
    2.2.1. The Dell System   9
    2.2.2. The British Steel System   10
    2.2.3. The Benetton System   10
    2.2.4. The Japanese Steel Industry   12
    2.2.5. The Chrysler System   14
2.3. Conclusion   14
    References   16

3
SUPPLY CHAIN AT THE STRATEGIC LEVEL   17

3.1. Introduction   17
3.2. Ideal Supply Chain Design   21
    3.2.1. Fundamentals   22
    3.2.2. Design Process   24
3.3. An Example of Mathematical Formulation   35
    3.3.1. Problem Formulation   36
    3.3.2. Management by Departments versus Supply Chain Approach   47
3.4. Dominant Partner   54
    3.4.1. Who Is the Dominant Partner?   54
    3.4.2. Why Is a Partner Dominant?   55
    3.4.3. Dominant Partners and Production Types   56
3.5. Conclusion   58
    References   58

4
SUPPLY CHAIN AT THE TACTICAL LEVEL   61

4.1. Introduction   61
4.2. Local Decisions and Global Consequences   62
4.3. Building an Efficient Supply Chain at the Tactical Level   65
    4.3.1. Tactical and Strategic Levels   65
    4.3.2. The Tactical Objectives in a Supply Chain   66
4.4. Performance Evaluation of a Supply Chain   86
    4.4.1. Financial Evaluation   87
    4.4.2. Operational Evaluation   89
4.5. Conclusion   93
    References   94

5
PRODUCT DEVELOPMENT IN A SUPPLY CHAIN   97

5.1. Introduction   97
5.2. Stages in the Product Life Cycle   99
    5.2.1. Product Identification   100
    5.2.2. Product Design and Development   103

    5.2.3. Product Introduction   114
    5.2.4. Product Sustenance   116
    5.2.5. Product Phaseout   118
5.3. Conclusion   120
    References   120

6

ENABLING TECHNOLOGIES   125

6.1. Technical Enablers   125
    6.1.1. Hardware Development   126
    6.1.2. Software Development   128
    6.1.3. Business Needs   132
    6.1.4. Human Resource Development   134
6.2. Application Enablers   135
    6.2.1. Legacy Applications   135
    6.2.2. Enterprise Resource Planning Applications   137
    6.2.3. Supply Chain Planning Applications   145
    6.2.4. Internet Business Applications   150
6.3. Conclusion   152
    References   152

7

CONCLUSION   155

APPENDIX A   159

A.1. Introduction   159
A.2. The Linear Manufacturing System   161
    A.2.1. Problem Setting   161
    A.2.2. An FBEST Algorithm   163
    A.2.3. Example   166
A.3. Generalization   168
    A.3.1. Notations and Problem Formulation   168
    A.3.2. Optimal Solution for the Acyclic Production System   170
    A.3.3. Illustrative Example   177
A.4. Work-in-Process Regulation   178
A.5. Conclusion   181
    References   181

Foreword

We are on the cusp of the next industrial revolution. We have entered the new millennium with the power of the Internet and information in our hands. The power of this new trend is ours to harness. We can revolutionize the way business is done and squeeze out the inefficiencies from our businesses. We can attain an order of magnitude increase in productivity of our resources. Manufacturing costs can be cut in half. It would be a pity not to use this opportunity to its fullest extent. The supply chain paradigm along with the enablers provided by the Internet technology can help manufacturing companies capitalize on this opportunity.

Early adopters of the supply chain paradigm in the manufacturing sector saw this potential in the 1990s and launched efforts in this direction. Their main focus was to improve the internal efficiencies of their organizations by driving down inventories, improving throughput, reducing manufacturing lead times, and increasing customer service levels. However, the paradigm is evolving, as are the level and scope of acceptance of this paradigm. The supply chain paradigm is no longer limited to being a tool for the logistics and operations group. It is gradually becoming the centerpiece of a company's strategy and involves the entire enterprise from R&D, engineering, and manufacturing to sales, marketing, and customer management.

The fundamental shift in the supply chain paradigm during the past couple of years has been the increased emphasis on improving outward-facing activities in an organization, such as customer management, relationships with suppliers and retailers, and the ability to do business over the Internet. This new paradigm of e-business requires fundamental changes to the decision-making process of an organization. Organizations have to tailor their decision-making process to have (i) increased emphasis on forward visibility into the relevant information needed for decision making; (ii) increased velocity in decision making to capitalize on the available opportunity; (iii) increased flexibility to cope with the changing environment; (iv) increased transparency in its operations to gain the confidence of its partners; and (v) knowledge of individual customers in order to more closely cater to their needs.

The latest development in the supply chain and e-business paradigm is the emergence of e-marketplaces. These virtual marketplaces on the Internet help businesses work more intelligently with their partners, suppliers, service suppliers, and customers to conduct business together in real time and to make more profitable decisions by effectively managing all their business processes, including procurement, fulfillment, product development, and customer care. E-marketplaces enable organizations to enhance their revenues by providing opportunities to (i) reach out to different customer segments; (ii) personalize their offerings to each customer; (iii) bundle products and services with complementary providers; and (iv) have greater velocity, flexibility, and transparency. E-marketplaces enable organizations to reduce their expenses by providing opportunities to (i) consolidate purchasing, (ii) exploit forward visibility and velocity to avoid inefficiencies, (iii) participate in auctions and bids to get the best price, and (iv) participate in marketplaces that provide value-added services and thereby help reduce expenses. E-marketplaces enable organizations to improve asset utilization by providing opportunities to (i) minimize the need for physical assets with better information, velocity, and intelligence in decision making, and (ii) providing a more effective utilization of the assets of all partners.

In this book, Govil and Proth highlight important concepts of supply chains that are very relevant in the current environment. They rightly identify customer satisfaction as the primary goal of a supply chain. Supply chains have to be passionate about fulfilling customer needs. They have to focus intimately on understanding customer needs because customers are the reason for their existence. The concept of a sharing mechanism presented in this book is fundamental to the success of the e-marketplace. To be successful, e-marketplaces have to be transparent and fair to all participants, independent of whether they are buyers or sellers. There can be no marketplace until both buyers and sellers have an incentive to participate in it, and a fair sharing mechanism is a strong incentive for participation.

This book approaches the supply chain from a broader perspective and touches on areas from product development to customer management. This book provides tools for designing supply chains and then effectively measuring their performance to ensure their success. It also highlights the need for an integrated information system throughout the entire supply chain to enable real-time information dissemination and decision making.

Currently, there is uncertainty and doubt in organizations regarding the strength and the opportunities that these new paradigms of supply chains, e-business, and e-marketplaces present. This environment represents a rare opportunity to those companies that will envision the opportunity and seize it with vigor. They can capitalize on the opportunity by having a positive attitude, adapting quickly to the conditions at hand, executing effectively, and winning decisively. Historically, whenever change is in the air, new leaders inevitably emerge. The new market leaders will have to actively shape the landscape that surrounds them. They will have to embrace e-business and profit from the e-marketplaces by moving faster than anyone else.

Sanjiv Sidhu
Chairman, i2 Technologies
Dallas, Texas

SUPPLY CHAIN DESIGN AND MANAGEMENT

1

INTRODUCTION

Developments in the field of production management since World War II have been limited to the improvement of activities related to production control and design in individual functional areas such as inventory management, planning and scheduling of manufacturing activities, modeling and evaluation of manufacturing systems, layout problems, group technology, system design approaches, and design and control of information flows, to quote only a few.

Despite competition among companies and an ever-changing market, the basic structure of manufacturing firms remained quite stable during this period. The most significant change was a tendency to increase automation, which affected neither the structure of the physical system (PS) nor the structure of the decision-making system (DMS). As a consequence, the problems researchers had to face did not evolve much during this period, except in the following aspects:

- The size of problems to be solved tended to increase due to increasing complexity of products as well as production systems.
- Time constraints became stronger due to competition.
- The number of tools available in production management exploded.

Although these tools were derived from a limited number of basic principles (MRP I, MRP II, TQM, JIT, etc.), the fact that they were developed independently from each other made a standardized approach quite impossible and thus handicapped the introduction of rationale in production management in most companies.

To summarize, the major economic changes that occurred in the 1950s first affected marketing strategies and then, in the 1980s and 1990s, affected production systems, calling for more automation without fundamentally perturbing the structures of DMSs and PSs.

Only recently have the pressure of the competitive market and new information technologies affected the structures of the production systems, calling for

- Integration of the activities that cover the whole spectrum of production from customers' requirements to products: This leads to a new DMS structure.
- Increasing flexibility of the PS structures by applying, among others, the concept of independent but correlated functional units. These units are managed independently from each other but receive information and are subject to constraints that guarantee that their activities converge toward the same goal.

The supply chain paradigm is a way to deal with this new situation. However, it seems difficult not only to define this paradigm as a whole but also to clearly specify a general process to implement it.

Numerous articles and books have been written on the subject, and conferences related to production management often include sessions on supply chains. Their goals have been to explain how and why globalization has resulted in new behavior of the people involved in this market (i.e., each one of us) and to identify these new behaviors. Sometimes, the literature proposes limited and qualitative models of supply chains that analyze a specific behavior of such systems.

The ideas presented in most of the books to date are of utmost interest. Some of them emphasize the importance of logistics and present supply chains as an extension of logistics management. Others consider that customers' perception of performances must be paramount and that, as a consequence, a successful supply chain is an organization that mainly focuses on improving the visibility of customers' demands and disseminating information among the participants to the supply chain. The importance of human working flexibility for a successful supply chain is outlined by some authors, whereas cost reduction is the most important objective for others.

Here, we summarize the approaches from some of the books currently available that present supply chains from similar, but differing, points of view. William C. Copacino (1997) presents an overview of the evolution of logistics and supply chain management. Throughout the seven chapters of this book, the reader is invited to follow the birth and the growth of this new paradigm in the business community. The story starts in the middle of the 1980s and ends in the middle of the 1990s.

M. Christopher (1998) claims that logistics and supply chain management are the two components of modern management, and that it is through these components that "the twin goals of cost reduction and service enhancement can be achieved." Note that in this approach, logistics is not a component of supply chains but a resource that makes it possible to implement a supply chain. This book is full of examples and suggestions and is one of the most interesting on the subject.

The book edited by John Gattorna (1998) is organized around the so-called Strategic Alignment Model. Gattorna claims that the ingredient that is missing in the current supply chain approaches is human behavior, which "generates and amplifies the pulses that reverberate through the supply chain." Thus, from his point of view, it is necessary to introduce the Strategic Alignment Model that "brings together the external market's dynamics, the firm's strategic response(s) and the firm's internal capability to execute this desired alignment, through the appropriate subcultures and leadership style(s) built into the organization." The papers brought together in this book examine not only how to tailor logistics and products to the needs of customers, meet customer satisfaction and demand, and develop winning collaborations with competitors, but also how to master complex channel dynamics, enhance supply chain decision making, win new customers, use cultural capability to improve supply chains, etc. This book insists on the importance of human behavior in the functioning and evolution of supply chains.

Charles H. Fine (1998) studies the influence of the choices made when designing supply chains on company performances. To do this, he restricts himself to the study of the fastest evolving companies (i.e., the companies with the greatest "clock speed"), the goal being to study as many companies as possible. The author observes these companies as a biologist observes fruit flies. He is interested in the migration of power and value up and down in the supply chain and the speed of these changes depending on the clock speed of the systems under consideration. He also implicitly considers the invariants existing in the structure of a supply chain since he draws general conclusions from the observation of fast clock speed companies. This hypothesis is questionable to us and, as we will show in this book, we prefer to help managers choose their own structure by suggesting different possibilities derived from the information about the system and checking the consistency of the choices.

Philip B. Schary and Tage Skjott-Larsen (1995) present an approach of supply chains that takes into account the specificities of European companies; however, they refer constantly to North American situations. For these authors, a supply chain includes all the processes that add value to material and bring the final product to the customer at the lowest price. This is a general formulation, which is completed by outlining that supply chain is more than logistics: It is a supraorganization that includes and coordinates all the activities that lead from raw material to final product. The authors point out that cooperative arrangements between the firms that comprise a supply chain are of utmost importance for its success, but no precise checklist of the kinds of arrangements that should be considered is provided.

The conclusion that can be drawn when reading the existing literature is somewhat disappointing. On the one hand, a huge number of exciting ideas and examples are proposed to the reader; on the other hand, there is a lack of information on adapting these ideas to specific situations.

We first consider strategic issues facing a supply chain. It is difficult to forecast the consequences of a strategic decision made at a given point of a system on other parts of the system, especially when the supply chain includes different, independent companies. For instance, the consequences of a strategic decision made at the production level (increasing production capacity by buying new resources and increasing working hours) on inventories, suppliers, customers, logistics, human resources, and marketing are difficult to forecast in terms of revenue and cost. Even the mechanism that disseminates such a decision among the supply chain is difficult to analyze for managers, and they are often surprised by the unexpected side effects that arise after making such a decision. When several companies are involved in the supply chain, a strategic decision made by one of the partners to improve its own performance may have a negative impact on other partners. It is well-known that the global optimum in a complex system is not the concatenation of local optima. Thus, it is necessary to include in any strategic supply chain model a "sharing process." Such a mechanism, which should be included in the design of any strategic supply chain model, is a process that adequately shares revenues and losses of the whole system among all the activities. It guarantees the consistency of the whole system by ensuring that the improvement to the efficiency of one of the partners is not at the cost of others.

Managers are calling for a toolbox that can help them design the strategic model of the supply chain in which they are involved. They want to mainly understand how their strategic decisions, and the ones of their partners, disseminate in the whole supply chain. They want to evaluate their commercial position in the supply chain by examining the rules that manage the sharing process and analyzing their consequences. They also want to use their strategic model to adapt their supply chain to the ever-changing market. Providing the basis of such a toolbox is the most important contribution of this book.

A definition of a supply chain is proposed in Chapter 2. The characteristics of such a system are also described in this chapter. Chapter 3 is dedicated to the objectives to be reached at the strategic level in order to meet the required characteristics. Chapter 4 discusses how the tactical-level decisions should follow from the strategic-level decisions and how that decision-making process should be designed for an effective supply chain. Chapter 5 discusses the role of product development strategy in the design of a supply chain. The current state of the art in technologies that can enable the practice of supply chain management is presented in Chapter 6. Chapter 7 presents our conclusions.

We deliberately do not consider supply chains as biologists who would like to analyze the existing systems of this type and extract their common characteristics and evaluate their behavior with regard to these characteristics. We want to take a step forward and propose methodologies and tools to help managers overcome the complexity of these systems.

Recently, major changes have occurred in production systems due to the pressures of the competitive market and new information technologies, calling for

- Integration of activities
- Increasing flexibility
- Drastic reduction of costs

Unfortunately, the conclusion that can be drawn when reading the existing literature is somewhat disappointing. On the one hand, many exciting ideas and examples are proposed to the reader; on the other hand, there is a lack of information to adapt these ideas to specific situations.

The goal of this book is to assimilate and complete information concerning the tactical level of supply chains and to introduce a new mechanism at the strategic level, the sharing mechanism, that guarantees a fair distribution of benefits and losses among all the partners and thus increases the motivation of each one of them.

REFERENCES

Christopher, M., *Logistics and Supply Chain Management. Strategies for Reducing Cost and Improving Service,* 2nd ed. Financial Time Management, London, 1998. [ISBN 0-273-63049-0]

Copacino, W. C., *Supply Chain Management. The Basics and Beyond,* APICS Series on Resource Management. St. Lucie Press, Boca Raton, FL, 1997.

Fine, C. H., *Clock Speed. Winning Industry Control in the Age of Temporary Advantage.* Parseus, Reading, MA, 1998. [Library of Congress Catalog No. 98-86945]

Gattorna, J. (Ed.), *Strategic Supply Chain Alignment. Best Practice in Supply Chain Management.* Gover, Aldershot, UK, 1998. [ISBN 0-566-07825-2]

Poirier, C. C., and Reiter, S. E., *Supply Chain Optimization. Building the Strongest Total Business Network.* Berret-Koehler, San Francisco, 1996.

Ross, D. F., *Competing through Supply Chain Management. Creating Market-Winning Strategies through Supply Chain Partnerships.* Kluwer, Dordrecht, 1999.

Schary, P. B., and Skjott-Larsen, T., *Managing the Global Supply Chain.* Munksgaard, Copenhagen, 1995. [ISBN 87-16-13278-5]

2

DEFINITION OF A SUPPLY CHAIN

2.1. GLOBAL DEFINITION

A supply chain is a global network of organizations that cooperate to improve the flows of material and information between suppliers and customers at the lowest cost and the highest speed. The objective of a supply chain is customer satisfaction.

The use of the term "network" suggests that the companies involved in a supply chain could not only be companies that perform complementary activities but also companies that compete to perform the same activities. The definition also states that this network of organizations is considered globally, and that the partners cooperate. This means that, viewed from the outside, a supply chain is a unique entity with, in particular, a unique strategy. To obtain such an integrated system that is fair for each one of its participants, an internal policy that specifies the relationships between the participants should be defined and implemented. The goal of this internal policy is to ensure that workloads, benefits, and losses are fairly shared among the participants, or else internal fighting for power may weaken the supply chain. We claim that if partners are allowed to make their own decisions freely, assuming that these decisions meet the internal policy accepted by all the partners, the efficiency of the supply chain can be dramatically improved. In this book, we call the implementation of the internal policy in a supply chain a "sharing process." This sharing process should be reconsidered periodically to take into account the changes that occur in the market. It is designed at the strategic level and applied at the tactical level.

In today's supply chains, the most powerful partner imposes the strategy of the system. In the car industry, for instance car makers dictate their conditions to suppliers. Computer makers such as IBM and Digital used to dominate their suppliers. Numerous other examples of supply chains that work on a master–slave basis could be presented. That landscape is now changing. In the computer industry, hardware makers such as IBM lost their dominance to software makers such as Microsoft. The ever-increasing use of electronics in cars may result in the shift of power to the electronics industry in the future. Irrespective of who dominates a supply chain, it is difficult to cite supply chains whose partners cooperate following a fair policy defined well in advance. We think that the introduction of sharing mechanisms will be the next step in the evolution of supply chains since it is a "win-win" strategy.

In a supply chain, the flow of information moves upstream, whereas the flow of material moves downstream. Information flows from customers to retailers, manufacturing companies, and logistics and raw material providers. It is the way production systems have been working for decades. The difference in a supply chain is that all the partners should be informed simultaneously, and the information they receive should be sufficient for them to make their own decisions. The idea of introducing a common information system that sends the customers' information simultaneously to all the partners according to their needs is a major characteristic of the next generation of supply chains.

Material flows downstream from suppliers of raw material or components to customers. As with information, the flow of material should be coordinated among all partners. This implies that activities should be coordinated upstream and downstream. We analyze the conditions necessary to reach this goal in Chapter 4.

The fact that these flows should move at the lowest cost and at the highest speed should be obvious. We will provide some approaches to evaluate cost and speed. Customer satisfaction is the ultimate goal to reach for success. We explain how to derive production objectives from this ultimate goal in Chapter 4.

In a supply chain, partners should be allowed to freely make their own decisions, assuming that these decisions meet the internal policy accepted by all the partners.

In a supply chain, the flow of information moves upstream, whereas the flow of material moves downstream.

2.2. EXAMPLES AND COUNTEREXAMPLES OF SUPPLY CHAINS

2.2.1. THE DELL SYSTEM

At the beginning of its life, Dell Computer was involved in a system in which the partners behaved on a master–slave basis. Dell depended on Intel and Microsoft upstream and on well-informed and careful customers downstream, and it had to compete with numerous computer companies, such as IBM, Compaq, Hewlett–Packard (H.P.) and Acer. This example is a network in which competition is open among the computer makers and in which competition is open for anyone to win the ever-changing leadership. A part of this network is represented in Fig. 2.1.

This system is referred to as a supply chain in numerous books and papers. In our opinion, it does not really fit with the definition of a supply chain as defined previously since its components do not cooperate, and there is no common information system that feeds computer makers with information from customers. Furthermore, there is no fair policy established by the

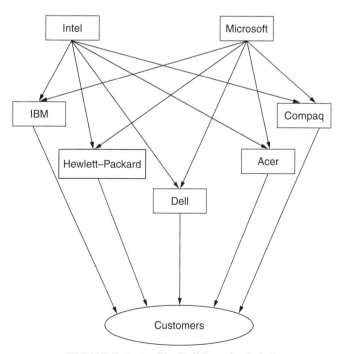

FIGURE 2.1 The Dell "supply chain."

competitors (IBM, H.P. Dell, Acer, and Compaq) to encourage cooperation: Only the rough market law prevails, and each computer maker has its own idea about what satisfies its customers. Thus, this system is simply a set of companies that need each other for doing business and that compete according to the free market rules.

2.2.2. THE BRITISH STEEL SYSTEM

Another example of a supply chain often mentioned in business books is British Steel (BS) as it was rebuilt during the 1990s. BS is one of the largest steel manufacturers in Europe and, as do any other steel manufacturers, provides steel to the automotive industry, which is one of its most important markets. The organization of BS can be schematized as follows:

- BS provides steel.
- British Steel Strip Products (BSSP), one of the operating businesses of BS, supplies material mainly to subcontractors, who process the steel to meet customers' requirements.
- British Steel Distribution and Service (BSD) aims at meeting the needs of its automotive industry customers and providing technical support to subcontractors from design to development. Several dedicated branches have been developed inside BSD to improve the efficiency of the system.

Figure 2.2 provides a simplified representation of this system. Indeed, this example is still not a supply chain if we refer to the definition presented at the beginning of this chapter since the sharing process is missing: It is replaced by the rules imposed by BS, the most powerful component of the system (i.e., the dominant partner). Nevertheless, assuming that these rules are designed taking into account the requirements of the subcontractors, this system approximately fits with the supply chain definition. It should be noted that, in this example, the leader (i.e., BS) has no competitor and distributes work to the other partners: These factors guarantee the stability of this system.

2.2.3. THE BENETTON SYSTEM

This system works typically on a master–slave basis. Benetton employs more than 800 subcontractors. Exclusivity is required from many of them. As a consequence, subcontractors cannot work for other customers in case of decreasing demand from Benetton. The way of doing business in Italy, where working rules are flexible and personal relationships are of utmost

2.2. EXAMPLES AND COUNTEREXAMPLES OF SUPPLY CHAINS

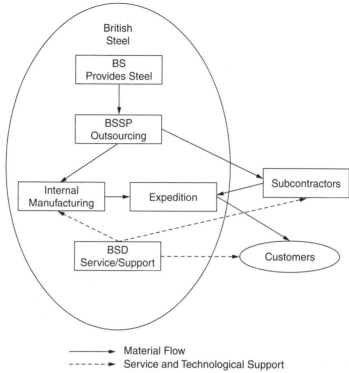

FIGURE 2.2 The British Steel supply chain.

importance, allows subcontractors to face these problems. Benetton outsources a large amount of the production to these subcontractors, the majority of which are located in the same region as Benetton (northeastern Italy).

Subcontractors are organized hierarchically from medium-size companies, which are the closest to Benetton, to family-owned firms located in low-wage regions of Italy or other countries, which often work for the high-level subcontractors. Benetton provides raw material and semifinished products to its subcontractors. It often generates the production plans of its closest subcontractors and acts as an advisor to them. It even takes care of risky investments by providing special machines that could become obsolete soon. Close relationships exist between the managers of Benetton and the subcontractors.

Only the core activities — that is, those that are responsible for the image of the company — remain in-house. These are mainly dyeing, quality management, cutting, and spinning. This organization makes it easy for

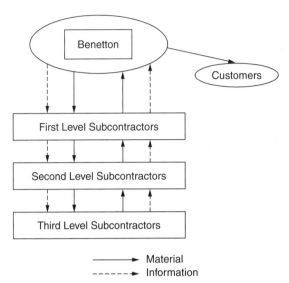

FIGURE 2.3 The Benetton system.

Benetton to increase or decrease its production capacity at a low cost, if needed, since many of the subcontractors depend exclusively on it. Due to the large number of subcontractors and the fact that many of these firms are composed of members of the same family, the risk of strikes against Benetton is negligible.

A positive aspect of Benetton's organization is that some of the activities could be moved to low-cost countries such as those in North Africa, Eastern Europe, or Asia, thereby reducing production costs. This system is represented in Fig. 2.3.

It is difficult to claim that Benetton's system is a supply chain. It is more an organization that manages to keep its workers outside the company in order to outsource problems due to changes of the market and to keep the benefits in-house.

This system does not really share risks. It does not work on a basis where all the partners set the cooperation rules. Furthermore, information is in the hands of Benetton alone.

2.2.4. THE JAPANESE STEEL INDUSTRY

The Japanese steel industry was in shambles at the end of World War II. Japan lacked natural resources and was forced to import most of the raw material and energy needed. Furthermore, due to their cultural background, the Japanese were reluctant to cooperate with foreign companies.

2.2. EXAMPLES AND COUNTEREXAMPLES OF SUPPLY CHAINS

To face this dramatic situation, Japanese managers developed a strategy that can be summarized as follows. A decision was made to expand business outside Japan and acquire mines and logistics resources abroad while carefully coordinating procurement and development of these means. The locations of steel plants were selected near ports in order to facilitate easy transportation and, thus, reduce transportation costs. Huge investments were made in ports and inland transportation. Logistics were integrated to remove the barriers in the flow of material between mines and users. Upstream, inputs were diversified to increase competition among suppliers and thus negotiate low costs. The strategy went even further by establishing minority ownership in some coal suppliers in order to influence the management of these companies in favor of the Japanese steel industry.

The Japanese trading companies have had a major role in the integration of the activities from raw material to steel. They procured raw material for the steel companies; acted as leaders in logistics, mining development, and warehousing; and managed the sale of finished products. This relationship is shown in Fig. 2.4.

To conclude, two main factors ensured the Japanese success. First, Japanese managers focused on the improvement of material flow by designing an efficient network (i.e., an efficient transportation system) with well-located nodes (i.e., mines, ports, and steel companies). Second, they took advantage of the trading companies that cemented the various activities of the system and thus played the role of leaders of the whole system, suggesting the rules that should be followed by all partner companies.

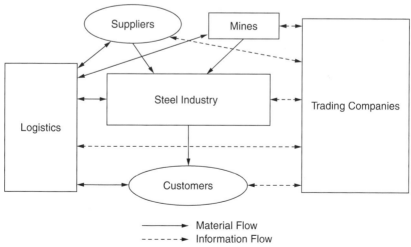

FIGURE 2.4 The Japanese steel industry.

This system exposes a real collaboration between the partners, who fought together to be competitive instead of fighting against each other to win the leadership of the system.

In addition, because information flowed easily in the whole system, the Japanese steel industry system is close to being a supply chain as defined at the beginning of this chapter, despite the fact that the collaboration with suppliers and customers is not as effective as it could be.

It should be noted that the reduction of costs was obtained not by short-sighted drastic measures, such as reduction in salaries, laying off workers and outsourcing parts, or passing on costs to suppliers. Instead, cost reduction was achieved through careful organization of the supply chain that removed inefficiency, reduced non-value-adding activities, and decreased costs for everyone involved.

2.2.5. THE CHRYSLER SYSTEM

Chrysler (now Daimler Chrysler) is one of the largest automobile manufacturers in the world. At the beginning of the 1990s, Chrysler was in financial trouble. At that point, the company decided to change the relationships with its suppliers. Previously, car makers use to outsource only those parts that required low technical knowledge, keeping complex parts and sophisticated subsystems in-house in order to control the entire know-how. Due to financial pressure, Chrysler decided to commit to long-term relationships with suppliers for developing parts and subsystems of a highly technical nature, sharing risks and benefits with suppliers. In doing so, Chrysler dramatically reduced design and development costs and encouraged suppliers to put Chrysler first in line among their clients. The Chrysler system of suppliers, which has since been copied by Ford and General Motors, has proved its efficiency.

The Chrysler system is not a supply chain by our definition since, in particular, a common information system and integrated logistics are not part of the system. However, this example shows how important and powerful a fair collaboration between partners is and how it can improve the efficiency and benefits of each one of the partners in the long term.

2.3. CONCLUSION

Even if the number of examples provided here were extended to infinity, the conclusion would still remain the same: A perfect supply chain does not exist. Some characteristics of supply chains have been developed due to the pressures of competition, but a global system fitting perfectly with the definition given at the beginning of this chapter is still a dream.

2.3. CONCLUSION

TABLE 2.1 Evaluation of the Systems

	Participants' spectrum	Cooperation	Material flow	Information flow	Cost	Customer satisfaction
Dell	5	2	10	2	10	10
British Steel	9	6	9	6	7	6
Benetton	5	1	7	2	9	5
Japanese steel industry	9	8	9	9	9	7
Chrysler	8	8	7	7	8	6

A perfect supply chain calls for a fair collaboration between all the participants, from suppliers to customers. This requires a clear definition of the collaboration rules between the partners, especially if they are competitors. Such a set of rules is called a sharing process, and it enables participants to share risks and benefits. Such a sharing process has to be defined during the design of the supply chain, long before conflicts occur among partners. It can be expressed in terms of contracts between all the partners, the goal of such contracts being to guarantee the best work practices among partners while allowing for healthy competition.

In Table 2.1, we give an evaluation (between 1 and 10) of the five systems presented in this chapter with regard to the following parameters:

- Participants' spectrum: Indicates the number of supply chain levels involved in the system. The greater the spectrum, the higher the score.
- Cooperation: The better the collaboration in terms of equal leadership of the partners, the greater the score.
- Material flow: The more important the part of the total material flow involved in the system, the greater the score.
- Information flow: The better the global information is shared among the different levels of the supply chain (from raw material to finished products), the greater the score.
- Cost: The higher the reduction of cost, the greater the score.
- Customer satisfaction: The higher customer satisfaction in the hierarchy of the criteria, the greater the score.

> None of the previous examples is a perfect supply chain, but each one of them shows at least one of the advantages of a perfect supply chain.

> Equal leadership of the partners can be observed in Chrysler and the Japanese steel industry.
>
> Dell is the best example of low work-in-process, cost reduction, and customer satisfaction.
>
> The Japanese steel industry is the best example of a system that encapsulates almost all components of a production spectrum and shares information among all the partners.

REFERENCES

Camagni, R., and Rabellotti, R., "Technology and organization in the Italian textile-clothing industry," *Entrepreneurship Regional Dev.* **4**, 271–285, 1992.

Dapiran, P., "Benetton—Global logistics in action," *Int. J. Physical Distribution Logistics* **22**(6), 7–11, 1992.

Dertouzos, M. L., Lester, R. K., and Solow, R. M., *Made in America: Regaining the Productive Edge.* MIT Press, Cambridge, MA, 1989.

Dyer, J. H., and Ouchi, W. G., "Japanese-style partnerships: Giving companies a competitive edge," *Sloan Management Rev.* **35**, 51–63, 1993.

Foster, T. A., "Global logistics Benetton style," *Distribution* **92**(10), 62–66, 1993.

Ketelhöhn, W., "An interview with Aldo Palmeri of Benetton: The early growth years," *Eur. Management J.* **11**(3), 321–331, 1993.

3

SUPPLY CHAIN AT THE STRATEGIC LEVEL

3.1. INTRODUCTION

The decisions at the strategic level of the supply chain lay out the framework of how the supply chain operates. The following five major activities take place within a supply chain at the strategic level:

- The **buy** activity includes the tasks of buying raw materials, components, resources, and services.
- The **make** activity concerns creating products or services as well as ensuring maintenance and repair of resources when needed and training workers — in sum, performing all the tasks that are needed for production.
- The **move** activity concerns transportation of materials and personnel inside and outside the supply chain.
- The **store** activity concerns the work-in-process (WIP) and raw material when it is waiting for transportation or transformation as well as the finished products waiting to be sent to customers.
- The **sell** activity concerns all the market-oriented activities, including marketing and sales.

Each of these activities is linked with all other activities and with the external world. As opposed to the day-to-day decisions in each of these activities, which are covered at the tactical level, the strategic level focuses on the long term. For example, the buy activity at the strategic level focuses on developing long-term relationships with suppliers. It is not shortsighted by the near-term goal of buying from the cheapest supplier. It would identify suppliers whose strategic goals are compatible with those of the supply chain. These goals include the following:

- Direction of technical innovation for the supplier
- Focus on quality
- Focus on reduction of cost
- Focus on reduction in response time

For example, an automobile manufacturer looking for a supplier of a fuel injection system for its engines would consider suppliers that have state-of-the-art technology and are moving toward greater integration of electronics into the system (which is a general trend for most automobile subsystems). The quality consciousness of the supplier and its commitment to short supply lead times would also be considered. Although important, lower cost would not be the only criteria. Also, there should be more than one supplier so that there is healthy competition, but there should not be too many suppliers because effective coordination will be lost.

The long-term issues for the make activity include decisions regarding the following:

- Improvement to the manufacturing processes.
- Investment and alignment of manufacturing resources along strategic goals (e.g., a chip manufacturer that considers wireless equipment the main focus of its strategy should invest more in equipment and human resources in that area).
- Management of production control should be aligned with the strategic goal; thus, a company may want to move from a make-to-stock policy to an assemble-to-order policy.
- Optimal portfolio planning of future products: A company should align its R&D and product development organization to reach the right market at the right time.
- Instilling the culture of being a world-class manufacturer with high emphasis on quality.

The long-term issues for the move activity require answering the following questions:

- How best to coordinate shipments between different locations? This includes exploring the possibility of negotiating contracts for consolidating shipments from different customers or even multiple

shipments from the same customer into one (if it does not delay the shipment and proves to be cheaper).
- What mode of transportation suits which customer?
- How to allocate resources between transportation equipment, such as trucks, and planes, ships and boats, and train containers?
- How can it provide more value-added service to its other partners; for example, can it provide packaging capability, can it act as a payment collector, and can it provide insurance?
- How should it keep its shipment tracking and planning systems state of the art?

The long-term issues for the store activity require answering the following questions:

- What storage facilities are needed, what size, and in what locations?
- How can the inventory be tracked in the most efficient and accurate way?
- What are the appropriate levels of inventories for the different products so as to achieve the desired level of fill rate at the target cost of storing?
- What are the ways to prevent losses during storage from damages and mishandling?
- How can the time for storing the products be reduced so that the inventory levels are at the lowest and hence not tying up much capital?

The long-term issues for the sell activity require answering the following questions:

- What markets to target?
- How, when, and how much to spend on marketing and promotion activities and how to allocate that among the different segments and channels?
- How to better anticipate customer requirements and how to meet them?
- How to respond to competitor activities?
- How to allocate resources to different sales channels, e.g., how much and what to sell on the Internet and how much and what to sell through the traditional stores?

- How to form alliances with other players in the market?
- How to coordinate and share information with the other participants of the supply chain to be effective in the marketplace?

The model encompassing the decision parameters and decision making that is associated with each of these individual activities will be referred to as a **module** in the remainder of this chapter. The set of modules along with their links constitute the model of the supply chain. Supply chain management links the previously mentioned activities, which convert raw materials into products and deliver those products to customers at the right time and at the right place in the most efficient way. In other words, supply chain management requires managing the flow of material from multiple sources to users and information from users to the components of the five activities mentioned previously.

Note that a change in any one of the activities influences the others, and an attempt to minimize any individual cost element may result in higher total cost. Some simple examples to illustrate this point are presented in Chapter 4. Furthermore, each activity included in a supply chain should be aligned with the financial objectives. In other words, the aim of a supply chain is not only to keep track of the consequences of a local decision on all the activities of the supply chain but also to evaluate the consequences of such a decision on revenue and cost.

The following considerations are of utmost importance in designing a supply chain at the strategic level:

- Analyze how the consequences of a local decision in one module would disseminate through the supply chain.
- Evaluate the financial implications of a local decision in one module on the other modules of the supply chain. This enables the computation of the consequence of a local decision on the total cost.
- Define how profit or loss would be shared between the different modules (sharing process).

The goal is to enable the partners to evaluate their own performance and, thus, make them as independent from each other as possible: This is a novelty compared to the traditional approaches.

The goal of this chapter is to focus on the strategic level of the supply chain and to provide information for developing a successful supply chain architecture.

In Section 3.2, we propose a design process that leads to an efficient supply chain architecture. The systematic approach proposed in this section is not how supply chains are currently designed because supply chains evolved from existing modules and dominant partners imposed their wishes on weaker partners (see the examples mentioned in Chapter 2). This aspect will be addressed in Section 3.4. In section 3.3, we present a sample mathematical model that highlights and contrasts the approach presented by us with the traditional approaches.

At the strategic level, a supply chain can be considered as being composed of five activities: **buy, make, move, store**, and **sell**. Each activity is a **module**. The set of modules along with their links constitute the model of the supply chain.

Designing the model of a supply chain at the strategic level requires analyzing how the consequences of a local decision in one module will disseminate through the supply chain, evaluating the financial implications of a local decision in one module on the other modules of the supply chain, and defining how profits or loss will be shared between the different modules (sharing process).

3.2. IDEAL SUPPLY CHAIN DESIGN

The architecture of the supply chain is the result of the resources gradually introduced for performing the projects that comprise the system. These projects correspond to the products and/or services that the system provides to its customers. A supply chain is not a frozen production system but a production system that can be extended or reduced, depending on the projects in progress, i.e., the number and volume of products/services being provided by the company at any given point in time. If the company is introducing a new product or the demand for an existing product (hence, its planned supply) is increasing, the company would need more resources. On the other hand, if the company is discontinuing a product or the demand of a product (hence, its planned supply) is decreasing, then the company would need less resources. In the remainder of this section, we assume that some projects are already in progress, and that a new project that consists of launching new types of products is considered. This project covers the five activities previously mentioned — **buy, make, move, store**, and **sell**. The parameters that are mentioned in this section concern only the new project. The capacities under consideration are those related to the new project (incremental capacities). Similarly, costs and benefits are incremental values.

3.2.1. FUNDAMENTALS

Each module, corresponding to one of the five major activities, is defined by

- Its state variables, whose values define the state of the module. For instance, some of these variables are related to the capacity of the supply chain for each operation that can be performed (**make** activity), the storage capacity for each type of WIP (**store** activity), or the customers with their buying capacity (**sell** activity).
- The set of feasible controls, which is the set of possible decisions that can be made regarding a module. Note that some of these decisions may or may not be feasible, depending on the constraints that apply.
- The events that may be generated by the external world, such as customers' requirements, changes in the prices of raw materials, and new competitors.
- The constraints applied on the control that are physical constraints (e.g., capacity) or constraints generated by other modules (the **sell** module can be constrained by the production capacity, constraints that apply to the **make** activity).
- The operational costs associated with each state of the module.

The modules considered at this stage are high-level modules. The set of these modules constitutes the high-level model of the supply chain. The goal of this model is to capture the interdependency of the different activities that constitute the supply chain and to evaluate the consequences of local decisions on the whole system. Evaluating the consequences of local decisions should be considered as the implementation of algorithms or checklists that will help managers to perform this type of evaluation. Note that the information required for building such a model is obtained from the expertise of managers and statistical data available from the company. The research conducted in the activity domain is another source of information.

3.2.1.1. An Example

Consider the module "make" as an example to illustrate the terms introduced in previous sections. In production systems, this module could be the high-level model of a manufacturing system.

3.2. IDEAL SUPPLY CHAIN DESIGN

The set of state variables may include variables assigned to the volumes of the different product families to be manufactured during the next n quarters and to the capacity of the system in terms of physical and human resources.

In each time period, the decisions that can be made concerning this module (i.e., the feasible control) are, for instance, a change in the production mix and/or a change in the production level. These decisions may require new production resources. They are constrained not only by the capacity of the system (usually the bottleneck resources and the workforce) but also by

- The raw materials available — this depends on the buy module.
- The possibility of selling (or not selling) the additional production quantities that may call for new advertising, new skills, or new maintenance management — this depends on the sell module.
- The possibility of introducing (or not introducing) new storage facilities and/or new storage management — this depends on the store module.
- The transportation capacity, which concerns the move module. Thus, a decision made in the make module may require new transportation resources.

Note that a decision made concerning the make activity influences the behavior of the other activities, which in turn modify the constraints that are applied on the make activity. This iterative aspect has to be incorporated in the design approach. New demands from customers and new strategies developed by competitors are common external events that can also affect the supply chain.

Finally, the operational cost of the make activity, the new operational costs of the other activities resulting from the change of their behavior due to the local decision made in the make activity, as well as the changes in the production revenues or losses resulting from the local decision should be evaluated. Furthermore, global revenues and losses should be shared among the activities.

3.2.1.2. Remark on the Components of a Supply Chain

A priori, the supply chain model to be designed should take into account two cases: (i) the case in which all the components belong to the same company and (ii) the case in which some of the components do not.

In the first case, the goal is obviously to increase the total profit generated by the system, even if the decisions to be made to reach this goal result in increasing the costs of some activities. For instance, it may be necessary to increase inventory levels, and thus to increase the cost associated with the store activity, to deal with competitors' strategies (i.e., to favor the sell activity). Similarly, it may be necessary to increase the cost of the buy activity to allow the make activity to deal with demand.

When different companies perform these activities, total benefit should be distributed at the points where flows (flows of information and flows of material) transit from one company to another. A transportation company (move activity) that is forced to increase the number of its transportation resources because the production at a partner company has increased (make activity) may increase its transportation costs, unless the increase in production generates an economy of scale large enough to compensate the additional costs.

An ideal supply chain design process should involve the following steps:

1. Identify and analyze the components (i.e., the modules) of the supply chain.
2. Analyze how the consequences of a local decision will disseminate through the supply chain in terms of constraints.
3. Analyze how the consequences of a local decision will disseminate through the supply chain in terms of costs and revenues.
4. Identify the external events that may influence the supply chain.
5. Analyze the possible consequences of these events on the five activities in terms of constraints.
6. Analyze the possible consequences of these events on the five activities in terms of costs and revenues.
7. Analyze how to share revenues and losses among the partners.

The last step is certainly the most difficult to introduce into the architecture of an ideal supply chain when partners are independent companies since dominant partners always tend to increase their benefits at the expense of weaker partners. We think that this attitude handicaps the whole supply chain in the medium term. The next section discusses the previously mentioned steps in detail.

3.2.2. DESIGN PROCESS

3.2.2.1. Identify and Analyze the Components of the Supply Chain

The tasks discussed in the following sections should be performed to design each one of the five modules of a supply chain.

3.2. IDEAL SUPPLY CHAIN DESIGN 25

Task A: Identify the Parameters That Characterize Each One of the Five Activities

For instance,

- The buy activity can be characterized by the list of raw materials that need to be bought, the list of the suppliers available for each raw material, the time to delivery, and the cost of the materials and their quality level.
- The make activity will probably be characterized by the production capacity related to some significant production mix and the related production cycle. At the strategic level, average setup times may also be part of the parameters, as well as the average production costs per unit for each possible mix.
- The move activity can be characterized by the transportation capacities related to different types of WIP between different significant pairs of points of the production system as well as the capacity related to transportation of finished goods between points of production, storage, and sale. The average transportation costs of a product unit of each type between the significant pairs of points also belong to the parameters of the move activity.
- The store activity can be characterized by the storage capacity. Another parameter is the average inventory cost per unit of product of each type and per unit of time at each storage level.
- The sell activity can be characterized by the buying capacities of the customers, the human resources available for performing the sell activity, the physical resources that are available, the average costs of the different types of actions available to promote a product, and the average selling costs of each type of product.

The previous examples are given only to provide insight into the definition of activities. In this task, designers also have to develop standards for information exchanges between modules so that information can be effectively shared by all partners in the supply chain to make informed business decisions.

In summary, the results of this study should include a list of parameters of importance according to the goal of the model and the possible values of these parameters.

A set of values associated with these parameters defines the state of the activity. Furthermore, the outputs to be identified for each module are

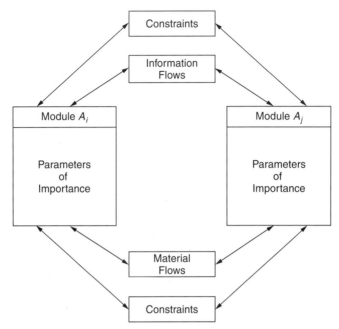

FIGURE 3.1   Modules and their connections.

- The operation cost
- Material flows toward other activities
- Information flows toward other activities
- The sharing process between the modules

These outputs are functions of the state of the activity under consideration.

Figure 3.1 summarizes the result of the identification of the modules that compose a supply chain.

Task B: Identify the Set of Decision Types That Can Be Made Regarding This Activity

An example has been previously proposed for the make activity. Most of the decisions can be expressed in the following terms:

- The type of decision — for instance, the raw materials to be bought (buy activity) and the mix to be manufactured (make activity).
- A numerical value, which usually represents a quantity: quantity to be bought (buy activity), quantity to be manufactured (make activity), quantity to be moved between a specific pair of points (move activity), etc. These values are part of the state of the system.

3.2. IDEAL SUPPLY CHAIN DESIGN

Task C: Identify the Set of Constraint Types That Apply to the Control

A constraint is an upper or lower limit that applies directly or indirectly to the quantities introduced previously. Constraints may or may not be independent from each other.

For instance, the minimal quantity of raw material a provider is prepared to deliver, the minimal delivery time of some components, and the maximal number of orders that can be managed each week due to the available raw materials are constraints that may apply to the buy activity.

Constraints that apply to the make activity include the average manpower available during each strategic elementary period (e.g., 1 month), the average availability of each type of physical resources during each strategic elementary period, and the minimal time period during which the same mix should be produced (setup constraint).

The constraints on the store activity include the level of the safety stocks of each type of product, which should be as low as possible, the average number of employees available, and the maximum value of the inventory of finished products in terms of tie-up of capital.

The constraints that apply to the move activity are the maximal number of employees and transportation resources available, the minimal transportation times between pairs of significant points, and the minimal and maximal loads of the transportation resources, etc.

The constraints on the sell activity include financial constraints such as the maximal marketing expenses, constraints on delivery times, and constraints on customer service levels and the number of employees available.

Remember that these constraints are global since we are at the strategic level.

Task D: Identify the Inputs and the Outputs of the Modules

Inputs and outputs of a module are material, products, and information. In some cases, financial flows should be added to these inputs and outputs. Financial flows are most relevant to the **buy** and **sell** modules as well as between modules that do not belong to the same company.

A decision made in an activity changes the state of the activity and changes the constraints imposed on the other activities, which in turn may call for decisions in these activities.

Figure 3.2 presents the consequences of a decision made in activity i, denoted by A_i, on A_i and another activity A_j. The model at the strategic level should lay down the rules on how the individual modules should react at the tactical level to the decisions in the other modules.

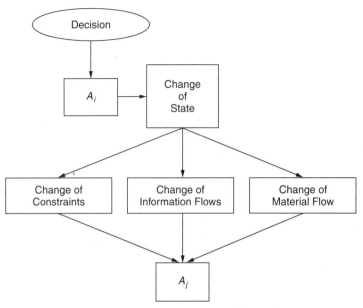

FIGURE 3.2 Consequences of a decision.

3.2.2.2. Dissemination of a Local Decision in Terms of Constraints

Task E: Analyze How the Constraints Are Modified by the Decisions Made on Other Activities

A systematic analysis should be conducted in this task to establish the relationship between a decision (defined by its type and the related numerical value) and the changes in the constraints that apply to the controls of the activities of the supply chain (i.e., the values of the upper and lower bounds). This analysis should lead to algorithms or checklists associated with each type of decision. Note that a change in the constraints that apply to the controls of other activities may require a change of these controls.

Consider the sell activity. The goal of this activity is to maximize the sales of the supply chain. The most common strategies applied for a short-term increase of sales are promotions. The level of promotions is tied to the expected increase in the sales volume of the products. The constraints imposed on the sell activity, in terms of how much they should spend on promotion during a period of time, are determined by the level of inventory in the store activity and the expected production volume during that period of time by the make activity.

3.2.2.3. Dissemination of a Local Decision in Terms of Costs and Revenues

Task F: Analyze the Financial Consequences of Each Type of Decision

The goal of this task is to establish for each decision type an algorithm or a checklist that will help managers to evaluate the cost or income resulting from a local decision of this type in each one of the activities of the supply chain. Each local decision leads to costs and influences revenues. The evaluation of costs and revenues is a key element of the design process. For each type of decision, the changes in cost and revenue are functions of the value of the parameters associated with the decision.

Reduction in inventory is an important objective of the supply chain. At the same time, customer satisfaction is also a major objective. In a make-to-stock type of environment, these two objectives could be conflicting. Reduction in inventory levels results in an immediate decrease in the capital tied up and, hence, improved cash flow. However, carrying this to the extreme may result in unfulfilled customer orders and, hence, not only reduction in revenue and profit in the short term but also loss of customer goodwill and market share in the long term.

3.2.2.4. Identify the External Events of Interest for the Supply Chain

Task G: Identification of the External Events

External events include unexpected demands, changes in the strategy of competitors, and new types of markets. Designers have to precisely identify external events that may happen in these domains for their specific supply chain.

The high-tech industry is very sensitive to changes in technology because technology can become obsolete very fast. Also, the market share can change overnight. Thus, these companies are very sensitive to any developments in technology in their domain and try to be proactive and devote a significant share of their profits to research and development. News of a new technology from a competitor often forces these companies to throw their development efforts into overdrive so as not to be left behind.

Similarly, marketing and promotion activities from competitors bring about a strong response from companies for whom the customer base is not stable and loyal.

3.2.2.5. Consequences of the External Events in Terms of Constraints

Task H: Analysis of the Consequences of External Events in Terms of Constraints on the Activities

Unexpected demand results in constraints on the capacities of the activities. For instance, if some new demand appears, one will have to adjust at least the constraints on the buy, make, store, and sell activities to deal with this demand. Note again that a change in the constraints that apply to the controls of other activities may require a change of these controls. Similarly, changes in the competitors' strategy may lead to changes in the types of products (**make** activity) and a new type of market may introduce severe constraints on each activity.

The goal of task H is to help managers precisely evaluate the new constraints that apply to the activities when an external event arises. This can be done by providing checklists or algorithms.

3.2.2.6. Consequences of the External Events in Terms of Costs and Revenues

Task I: Analysis of Financial Consequences of External Events

This task is based on the results of task H. It consists of deriving the changes in the costs in each one of the activities from the changes in the constraints. This should be made through checklists or algorithms flexible enough to integrate the expertise of the managers.

Thus, if the market share can be correlated to the level of marketing expenditure by a company, then an increase in marketing spending by a competitor has to be matched by the marketing spending of the company, depending on the available financial resources and the desired market share.

The introduction of a new product by a competitor should be evaluated in terms of its impact on the market share in the market segment of interest. This financial analysis should then be compared against the cost involved in designing, producing, and selling a competing product. Only if the returns are more than the cost should the company consider bringing the new product to the market. The company may be willing to lose that market share for the near term if it believes it has a product in the pipeline with better technology and market appeal that will help it regain its market share.

3.2.2.7. Sharing Revenues and Losses

Task J: The Sharing Process

The sharing process should facilitate the sharing of profit and loss of the whole system among all the activities so that a high efficiency of each one of the activities leads to a high efficiency of the whole system, with the efficiency of the system measured in terms of profit.

As mentioned previously, implementing a sharing process when some of the modules of the supply chain belong to different companies has never effectively occurred in supply chain history. In our opinion, this handicaps the efficient functioning of the supply chain.

The sharing process should be based on a set of common rules established by the managers in charge of the activities. The sharing should occur at the interfaces between organizations. There are two distinct interfaces between the organizations: the interface at which information is exchanged (e.g., the order placement and invoicing activity) and the interface at which materials and services are exchanged (e.g., delivery of raw material by the supplier to the activity in charge of transportation and the consequent delivery of that material to the make activity).

Consequently, there should be two distinct sets of rules for these types of sharing. Thus, the sharing process includes the following:

- The information that must be shared by the partners. This includes the strategic decisions made in each activity, the results obtained by the activity, and global information on the state of the activity. This point is of utmost importance when different companies perform the activities that comprise the supply chain.
- The sharing of benefits and losses among partners. The mechanism used to share benefits and losses is not unique and should be decided by the partners when the supply chain is designed. The aim of this sharing process should be to evenly distribute the potentially high cost of one activity (incurred keeping strategic objectives in mind) among all the constituents of the supply chain so as to pass on the benefits resulting from the high cost to the partner concerned. It seems that the best solution is to introduce one sharing process for each one of the projects performed by the supply chain.

The concept of a sharing process attacks at the roots of the inefficiencies of the supply chain, which are the culture of having a sense of control (or power) over other organizations, and the performance measurement systems of the organizations involved in the supply chain.

The main barrier to information sharing between the constituents of the supply chain is the fact that it is seen as sharing of power or of losing control. Some of the information that pertains to the core competencies of the organizations needs to be guarded for competitiveness. However, most information, especially that related to timely fulfillment of customer orders, should be made available to all partners as soon as possible and poses little threat to the competitiveness of any particular partner of the supply chain.

The sharing of benefits (or losses) between constituents of the supply chain (whether internal or external) is hampered by the performance measurement metrics in place for these constituents. For example, the buy activity is often measured by the cost of acquiring goods and services. Often, this becomes their sole criteria in evaluating suppliers and interacting with them. They try to drive the price of the bought commodity or service to the minimum and give preference to the supplier with the lowest cost. This narrow mind-set can in fact prove to be very costly in the long term for the supply chain. The lowest cost supplier may not provide the best quality and on-time delivery, leading to huge costs in terms of late delivery to customers as well as scrapping of the final product. Even if the company can force the supplier to supply high quality and on-time delivery at very low cost, this is not a sustainable solution. In the short term, the supplier may be able to do so by compromising its long-term goals such as investment in R&D and the technology provided by the vendor may become obsolete very soon. If the contract is awarded to a single supplier offering the lowest cost, the supply chain may face the prospect of increased prices from the supplier in the future since it is solely dependent on that supplier. However, a narrow attitude of buying everything from a different vendor with lowest individual cost of a component may result in a proliferation of suppliers, leading to high transaction expenses. A more strategic approach is to develop more meaningful relationships with the suppliers. Thus, instead of using cost as the sole decision criteria, the capability of the supplier in terms of quality, on-time performance, and dedication to innovation should also be considered. A slightly higher price for a commodity can be more than compensated by high quality and on-time delivery. Also, if the supplier is innovative, the supply chain can bring its own new products to market faster with the latest technology. The supply chain may invest in the R&D activity of the supplier and even outsource some design work to them. Instead of buying different components from many vendors, the supply chain may buy the components assembled in subsystems from a few vendors. This eliminates not only transactions with many different suppliers but also the need for in-house activity that adds little value to the product and is not the core competency of the supply chain. The supply

3.2. IDEAL SUPPLY CHAIN DESIGN

chain should also negotiate the terms and costs of special orders in advance, recognizing that special orders cost more. This would ensure that the orders from the supply chain receive the due attention of the supplier. This sharing process will be beneficial for both the suppliers and the supply chain.

The make activity is often measured by the utilization of its resources and its productivity in terms of the number of units produced. One of the main factors that reduces productivity is the time lost in setups when switching from the production of one type of product to that of another. Therefore, operational managers have a tendency to reduce the number of setups. Thus, they produce large numbers of units of the same type of product. This results in double loss for the supply chain. On the one hand, the required product is not made on time; on the other hand, huge quantities of some particular product are produced that consume scarce material and resources and have to sit in the inventory tying up capital and are prone to damage and obsolescence. Thus, the make activity should be compensated for its loss of productivity when the right products are made, at the right time, even if it means more setups. This "sharing" encourages better alignment of the objective of the make activity with the supply chain objective of customer satisfaction.

The store activity is often measured by the dollar value of the inventory. The natural incentive for the store activity is to reduce the level of inventory as much as possible. Inventory can be reduced by having a good estimate of expected demand of different products over time. The need for a high level of inventories is driven by two main factors: (i) the inability of the sell activity to precisely predict the demand for different products over time and (ii) the inability of the production system to meet demand in a small period of time. Therefore, these two activities should compensate the store activity for the high level of inventories.

The move activity is measured by on-time delivery and delivery cost. These are often conflicting objectives. More frequent deliveries often require transporting goods in quantities less than the full capacity of the equipment, (i.e., truck shipments with less than a truckload). Other activities of the supply chain can help the move activity meet these conflicting activities in two ways: (i) They can provide information about multiple shipments of small quantities in time for the move activity to consolidate those shipments and hence save cost and (ii) they can share the benefits achieved by more frequent shipments by paying slightly higher rates.

The sell activity is often measured by sales volume. Thus, there is an incentive to increase the sales revenue. The goal of the supply chain is higher customer satisfaction and at the same time profitability. The sales activity tries to increase sales either through selling a higher quantity of some low-value product or through selling a higher priced product. Profit margins are often not of concern. Emphasis should be placed on more profitable products, and if this means lower revenue, the performance metrics of the sales activity should be compensated accordingly. Another example of conflicting interests is promotions by the sell activity that are not coordinated with the make and store activities. To increase the volume of sales, often promotions or discounts on products are offered. The sell activity should inform the make and store activities of this decision in advance so that they produce and store the right product in the right quantities for the event. The make activity should be compensated for any additional costs including overtime and the store activity should be compensated for the higher level of inventories for this period by the sell activity.

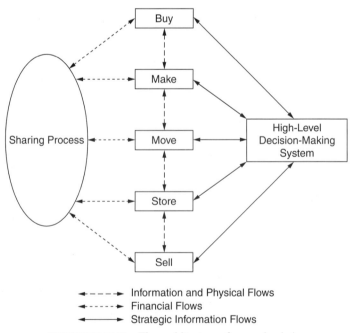

FIGURE 3.3   The architecture of a supply chain.

The ultimate goal of the sharing process is to motivate partners by linking their revenues to their efficiency. Figure 3.3 illustrates the architecture of a supply chain that may include different companies.

> Activities that lead from customers' requirement to customer satisfaction are considered as being independent from each other in terms of management. Nevertheless, they are synchronized and coordinated in a way that guarantees that efficient management of each of the activities leads to efficiency of the entire system. To reach this goal, one should (i) find out how a decision made in one of the activities disseminates in the other activities in terms of costs, and (ii) define a process to share revenues and losses of the whole system among all the activities so that high efficiency of each of the activities leads to high efficiency of the whole system, with efficiency being measured in terms of amount of benefit. This process is called the sharing process throughout the rest of the book.

3.3. AN EXAMPLE OF MATHEMATICAL FORMULATION

The goal of this section is to show how the decision-making process for the implementation of a new project, the introduction of a new product, in an existing supply chain can be expressed, at least partially, in mathematical terms. This project covers the five activities already mentioned — that is **buy**, **make**, **move**, **store**, and **sell**. Since the goal is to decide whether the new project should be implemented, we are at the strategic level in the decision-making process. Indeed, the model proposed here is particularly simple, but the process used to build this model can be easily generalized to make strategic decisions.

In the example presented here, we assume that customers are located in various regions of a country (or several countries). The global demand in each region is known and represented by a normal random variable. This demand has been estimated based on the techniques available in the supply chain and the goal is to meet this expected demand for the new product while maximizing the profits. Several retailers are candidates for participating in the new project in each region. Furthermore, several manufacturing units are candidates for supplying the retailers. In turn, several suppliers may feed manufacturing units.

The goals of this section are to
- Present the project. The presentation will be made activity by activity. In each activity, preliminary actions are required to provide data to the mathematical models that generate the final decisions.
- Use the mathematical models to explain the basic difference between a decision-making process that is functional and department oriented versus a supply chain-oriented decision-making process.

3.3.1. PROBLEM FORMULATION

We consider the five activities in the order in which the information flows in the system — that is, from customer demand to the suppliers of raw materials.

3.3.1.1. The Sell Activity

Implementing the mathematical model requires a preliminary activity that consists of selecting the subset of most likely retailers from the list of all available retailers from which a further "mathematical" selection will be made in each region in order to maximize the profit (or to minimize the loss). This preliminary selection is made taking into account the efficiency of the retailers, their dedication to quality, their financial and technical soundness, their ability to take responsibility for the new project, their willingness to participate in the common success of the supply chain by sharing information and know-how, their location, and so on. This preliminary selection activity also establishes the values of the parameters used in the mathematical model, such as the capacities of the retailers, the sum of the salaries of the employees assigned to the project at each retailer, the incremental marketing expenses (i.e., the additional marketing expenses made to promote the project), and the expenses incurred for servicing the customers.

The mathematical model presented here consists of defining the best repartition of available supply among the retailers of each region in order to meet the demands expressed in the regions and not to exceed the selling capacity of each retailer.

We consider r regions that comprise a given market. Each region $i \in \{1, 2, \ldots, r\}$ is supposed to consume d_i products of the type under consideration during a so-called elementary period that could be 3, 6, or more months. d_i is a normal random variable whose mean value is $\bar{d}_i$ and whose standard deviation is σ_i. In each region i, z_i retailers are prepared to sell

3.3. AN EXAMPLE OF MATHEMATICAL FORMULATION

the product. At the level of the sell activity, the goal is to define the quantity d_{ij} of product that should be sold by retailer $j \in \{1, 2, \ldots, z_i\}$ of region $i \in \{1, 2, \ldots, r\}$ during each elementary period. Quantities d_{ij} are also normal random variables whose mean values are denoted by $\bar{d}_{ij}$ and whose standard deviations are denoted by σ_{ij}. Thus,

$$\left. \begin{array}{l} \bar{d}_i = \sum_{j=1}^{z_i} \bar{d}_{ij} \\ \sigma_i = \sqrt{\sum_{j=1}^{z_i} \sigma_{ij}^2} \end{array} \right\}, \quad i = 1, 2, \ldots, r \qquad (3.1)$$

Also,

$$\sigma_{ij} = \sqrt{\frac{\bar{d}_{ij}}{\bar{d}_i}} \sigma_i \qquad (3.2)$$

We also know the maximal quantity D_{ij}^* of products that can be sold by retailer j of region i:

$$\bar{d}_{ij} \leq D_{ij}^*, \quad i = 1, 2, \ldots, r; \ j = 1, 2, \ldots, z_i \qquad (3.3)$$

and the profit (or the loss) b_{ij} made by retailer j of region i when selling one unit of product. This profit (or loss) is defined as the difference between the revenue and the marginal cost associated with the sell activity for the new project. This cost is the sum of the salaries of the employees specifically assigned to the project, the expenses incurred promoting the project (marketing), the expenses incurred implementing aftersale services, and so on.

The total average profit on T elementary periods, taking into account the interest profit α, is

$$\mathrm{BN}(\{\bar{d}_{ij}\}_{i=1,2,\ldots,r;\, j=1,2,\ldots,z_i}, T)$$

$$= \begin{cases} \dfrac{(1+\alpha)^T - 1}{\alpha} \sum_{i=1}^{r} \sum_{j=1}^{z_i} b_{ij} \cdot \bar{d}_{ij} & \text{if } \alpha > 0 \\ T \cdot \sum_{i=1}^{r} \sum_{j=1}^{z_i} b_{ij} \cdot \bar{d}_{ij} & \text{if } \alpha = 0 \end{cases} \qquad (\mathrm{F1})$$

where α is the interest rate per elementary period the supply chain would receive if the benefit were invested as cash at prevailing bank rates.

In Eq. (F1), we assume that the profit made during a given elementary period is available at the end of the period. The objective in Eq. (F1) is to maximize the profit (i.e., BN). The parameters to compute are the $\bar{d}_{ij}$ that must satisfy Eqs. (3.1) and (3.3).

Here, we have considered the relationship between the customers' demand in each region, denoted by d_i, and the average quantities $\bar{d}_{ij}$ to deliver to each retailer j from each region i. The objective under consideration is the profit or loss resulting from the assignment of finished products to retailers. The horizon T of the evaluation is given. It will be shown later that the parameters $\bar{d}_{ij}$ and the standard deviation σ_{ij} impose constraints on the move and store activities, which in turn impose constraints on the make and buy activities.

3.3.1.2. The Move Activity

The move activity includes all the transportation activities of the supply chain. In this example, we assume that the transportation activities are limited to the transportation of parts within the manufacturing units and from manufacturing units to retailers where they are stored. We assume that there is a one-to-one correspondence between the retailers and the storage facilities for the sake of simplicity. The goal is to minimize the sum of the investments required by new transportation resources and the amount of money needed for running the transportation resources in T elementary periods.

A preliminary selection of the vehicles used inside the manufacturing units and the transportation resources used to move parts from manufacturing units to retailers is made. The criteria considered to make the preliminary selection are related to the prices of the resources, their reliability, the maintenance costs, the payment facilities offered by the makers of the transportation resources, and so on.

Data obtained from this preliminary study include the capacities of the transportation resources selected, their maintenance costs, their reliability, the subsequent average costs for transporting one unit of product while it is inside the manufacturing units, and the production capacities of the manufacturing units. These data are used in the following mathematical optimization model that leads to the transportation flows that minimize the total transportation cost.

m manufacturing units are available to produce the $\bar{d} = \sum_{i=1}^{r} \bar{d}_i$ products required on the average during one elementary period. We denote by q_{kij} the quantity produced during each elementary period by manufacturing unit $k \in \{1, 2, \ldots, m\}$ for retailer j in region i.

We know the following:

- The transportation capacity q_{kij}^0 already available in the supply chain between manufacturing unit k and retailer j of region i.
- The incremental cost c_{kij} of the transportation of one unit of product from manufacturing unit k to retailer j of region i: This value is statistically derived. This includes the salaries of the

drivers, the maintenance costs of the trucks, the fuel costs, and, in general, all the expenses specifically connected with the new project.
- The capacity KT of a truck used to transport products from manufacturing units to retailers. We assume that all the trucks are the same.
- The cost C of a truck.
- The transportation capacity Q_k^0 already available in manufacturing unit k.
- The incremental cost s_k required for transporting one unit of product inside the manufacturing unit k. This value is statistical and established by similarity with other existing products.
- The capacity KT_k of a transportation unit that operates in manufacturing unit k.
- The cost C_k of a transportation unit that operates in manufacturing unit k.
- The upper bound on the production of the new product type in manufacturing unit k. This upper bound is denoted by p_k.

The following constraints hold:

$$\sum_{k=1}^{m} q_{kij} = \bar{d}_{ij}; \quad i = 1, 2, \ldots, r; \ j = 1, 2, \ldots, z_i \qquad (3.4)$$

Constraints in Eq. (3.4) mean that the total quantity of products that are transported from the manufacturing units to retailer j of region i during one elementary period is the mean value of the quantity that is sold by the retailer during the same period.

$$\sum_{i=1}^{r} \sum_{j=1}^{z_i} q_{kij} = \Omega_k \leq p_k; \quad i = 1, 2, \ldots, m \qquad (3.5)$$

Constraints in Eq. (3.5) mean that the quantity produced during one elementary period by a manufacturing unit must be less than its capacity during the same period.

A solution may exist only if the total production required during one elementary period is less than the total capacity of the manufacturing units, that is,

$$\sum_{k=1}^{m} p_k \geq \sum_{i=1}^{r} \bar{d}_i = \bar{d}$$

The total cost associated with the move activity, taking into account the interest rate, is

$$\mathrm{CM}(\{q_{kij}\}_{k=1,\ldots,m; i=1,\ldots,r; j=1,\ldots,z_i}, T)$$

$$= (1+\alpha)^T \cdot \sum_{k=1}^{m} \sum_{i=1}^{r} \sum_{j=1}^{z_i} C \left\lceil \frac{q_{kij} - q_{kij}^0}{KT} \right\rceil^+$$

$$+ \frac{(1+\alpha)^{T+1} - (1+\alpha)}{\alpha} \sum_{k=1}^{m} \sum_{i=1}^{r} \sum_{j=1}^{z_i} c_{kij} \cdot q_{kij}$$

$$+ (1+\alpha)^T \cdot \sum_{k=1}^{m} C_k \left\lceil \frac{\Omega_k - Q_k^0}{KT_k} \right\rceil^+$$

$$+ \frac{(1+\alpha)^{T+1} - (1+\alpha)}{\alpha} \cdot \sum_{k=1}^{m} [s_k \cdot \Omega_k] \qquad \text{(F2)}$$

In Eq. (F2) we assume that the costs incurred during one elementary period are incurred at the beginning of the period.

The variables whose values must be computed are q_{kij}. Expression (F2) is the sum of four terms that take into account the interest rate on each elementary period. The first term is the cost of the new trucks required (if any) for the new project. The second term is an evaluation of the running cost of the move activity between the manufacturing units and the retailers over T periods. The third term is the cost of the new transportation resources required (if any) for the new project. These resources are the ones that operate inside the manufacturing units. The last term is an evaluation of the running cost of the move activity inside the manufacturing units over T elementary periods.

In Eq. (F2), $\lceil a \rceil$ is the smallest integer greater than or equal to a, and $\alpha \in [0, 1]$ is the interest rate. T is the number of elementary periods under consideration. In the evaluation of a new project, T is the first period in which the sum of the benefits is supposed to exceed the sum of the costs and investment made for the project.

In the move activity, the goal is to link the quantities $\bar{d}_{ij}$ received by retailers during one elementary period and the quantities q_{kij} carried from manufacturing unit k to retailer j of region i during one elementary period while minimizing the total cost CM. The criteria under consideration include the operational cost of the transportation systems and the purchasing costs of the transportation resources required to implement the new project.

3.3.1.3. The Store Activity

The store activity usually includes the storage of WIP and finished products. A preliminary analysis is carried out for selecting storage resources that will be bought, if required, to complement the ones that are already available in the supply chain to store the new product. This selection is made taking into account the prices of the resources, the cost of their maintenance, the number of employees required to handle the products, the adequacy of the resources with regard to quality, etc.

As a result, this preliminary analysis provides the prices of the selected resources, all the costs associated with the use of these resources (salaries, maintenance, energy, if any). Another preliminary activity consists of defining the running cost of the existing storage facilities that will be used for the new project. The preliminary analysis also provides the costs that are used in the mathematical model presented here as well as the maximal storage capacities of each retailer.

In this example, we consider only the storage at the retailers' level. In other words, we assume that there is one-to-one correspondence between retailers and storage facilities (the retailers possibly own their own storage facility at their site) and no significant breakdowns happen at the manufacturing or the transportation levels. Only the randomness of the demand is significant enough to create inventories. We thus assume that the same quantity q_{kij} is transported during each elementary period from manufacturing unit $k \in \{1, 2, \ldots, m\}$ to retailer $j \in \{1, 2, \ldots, z_i\}$ of region $i \in \{1, 2, \ldots, r\}$ and that the demand d_{ij} that retailer j of region i should meet is a normal random variable as defined in Section 3.3.1.2.

Let v_{ij} be the cost required for storing one unit of product during one elementary period at the retailer j of region i. Similarly, h_{ij} denotes the cost incurred when one unit of product cannot be provided to customers during one elementary period at the retailer j of region i. y_{ij} is the target inventory level at the retailer j of region i. The target inventory level is supposed to provide an acceptable trade-off between inventory cost and the cost incurred when stock is running out.

The total inventory cost during one elementary period, assuming that y_{ij} is given, is computed as follows:

$$\text{CS}_{ij}(y_{ij}, 1)$$
$$= \frac{v_{ij}}{\sigma_{ij}\sqrt{2\pi}} \int_{d_{ij}=0}^{y_{ij}} \exp\left[-\frac{1}{2}\left(\frac{d_{ij} - \bar{d}_{ij}}{\sigma_{ij}}\right)^2\right] \cdot (-d_{ij} + y_{ij}) d(d_{ij})$$
$$+ \frac{h_{ij}}{\sigma_{ij}\sqrt{2\pi}} \int_{d_{ij}=y_{ij}}^{+\infty} \exp\left[-\frac{1}{2}\left(\frac{d_{ij} - \bar{d}_{ij}}{\sigma_{ij}}\right)^2\right] \cdot (d_{ij} - y_{ij}) d(d_{ij})$$

where d_{ij} is the random demand at retailer j of region i, and $\bar{d}_{ij}$ is the average value of d_{ij} and the constant replenishment of the inventory at the retailer j of region i. In this formulation, we consider that when it is impossible to meet a demand, the demand is delayed until the inventory contains enough products to meet it.

Assuming that all the safety stocks are given, the average total running cost during one elementary period is

$$CS(\{y_{ij}\}_{i=1,2,\ldots,r; j=1,2,\ldots,z_i}, 1) = \sum_{i=1}^{r} \sum_{j=1}^{z_i} CS_{ij}(y_{ij}, 1)$$

Assuming that all the safety stocks are given, the average total running cost during T elementary period is

$$CS(\{y_{ij}\}_{i=1,2,\ldots,r; j=1,2,\ldots,z_i}, T)$$
$$= \frac{(1+\alpha)^{T+1} - (1+\alpha)}{\alpha} CS(\{y_{ij}\}_{i=1,\ldots,r; j=1,\ldots,z_i}, 1)$$

The purchasing cost of one unit of storage capacity for retailer j in region i is denoted by f_{ij}. The storage capacity already available to retailer j of region i is denoted by L_{ij}, and the additional capacity to be purchased is denoted by R_{ij}. Finally, the total cost over the horizon T is

$$CS(T) = \frac{(1+\alpha)^{T+1} - (1+\alpha)}{\alpha} CS(\{y_{ij}\}_{i=1,\ldots,r; j=1,\ldots,z_i}, 1)$$
$$+ (1+\alpha)^T \sum_{i=1}^{r} \sum_{j=1}^{z_i} f_{ij} \cdot R_{ij} \qquad (F3)$$

The target inventory level is limited by the capacity of the inventory facilities. In this case,

$$y_{ij} + y_{ij}^0 = L_{ij} + R_{ij}, \quad i = 1, \ldots, r; \; j = 1, \ldots, z_i \qquad (3.6)$$

where y_{ij}^0 are the target inventory levels required for the projects already implemented. The variables whose value must be computed y_{ij} and R_{ij}.

In this model, since the goal is to find the optimal target inventory and since production is constant, Eq. (F3) can be optimized locally. In other words, minimizing Eq. (F3) under constraint (3.6) will lead to a part of the optimal solution.

This section was devoted to the relationship between the quantities of products sold by each retailer j from each region i and the safety stock y_{ij} to be kept by each retailer in order to meet customer demand with a given probability. The criterion under consideration is the sum of the inventory costs at retailer locations.

3.3. AN EXAMPLE OF MATHEMATICAL FORMULATION

3.3.1.4. The Make Activity

A preliminary task at this level consists of selecting the manufacturing units that will be used in the project. This selection is made taking into account, for each manufacturing unit, the prices of the resources to be bought, the running costs (salaries, maintenance, and energy consumption) of the resources that will be used for the new project, and their capacities. These data are required to feed the mathematical model presented later.

As mentioned earlier, m manufacturing units are available to meet customer demands. For each manufacturing unit $k \in \{1, 2, \ldots, m\}$, we know

- The capacity $\varphi^0_{k,g}$, $g \in \{1, 2, \ldots, G_k\}$ already available for resource type g of manufacturing unit k to manufacture the new type of products.
- The incremental cost λ_{kg} for performing the required operation on one unit of product using a resource of type g of manufacturing unit k.
- The number ω_{kg} of operations that can be performed by a resource of type g in manufacturing unit k during one elementary period.
- The cost Λ_{kg} of one resource of type g in manufacturing unit k.
- The quantity Ω_k to be manufactured by manufacturing unit k during one elementary period. These quantities are the variables that have to be defined.
- The upper bound p_k of the production of manufacturing unit k during one elementary period. This parameter was defined in Section 3.3.1.2.

The problem consists of assigning a production level Ω_k to each manufacturing unit. If the production level is assigned, then the total production cost is given by

$$\text{MAN}(\{\Omega_k\}_{k=1,\ldots,m}, T) = (1+\alpha)^T \sum_{k=1}^{m} \sum_{g=1}^{G_k} \lambda_{kg} \cdot \Omega_k$$

$$+ \frac{(1+\alpha)^{T+1} - (1+\alpha)}{\alpha} \sum_{k=1}^{m} \sum_{g=1}^{G_k} \Lambda_{kg}$$

$$\cdot \left\lceil \frac{\Omega_k - \varphi^0_{kg}}{\omega_{kg}} \right\rceil \quad \text{(F4)}$$

The following constraints hold:

$$\Omega_k \leq p_k \quad \text{for } k = 1, 2, \ldots, m \tag{3.7}$$

$$\sum_{k=1}^{m} \Omega_k = \sum_{i=1}^{r} \bar{d}_i \tag{3.8}$$

Constraint (3.7) expresses that the production of each manufacturing unit cannot exceed its capacity. Constraint (3.8) guarantees that the total production of the manufacturing units during one elementary period equals the mean value of the total demand.

The make activity establishes the relationship between the mean total demand and the total production during one elementary period. The criteria to take into account over the horizon T are the purchasing cost of the new resources required to perform the new project and the operational costs of the manufacturing units.

3.3.1.5. The Buy Activity

The goal of this activity is to provide raw material and components to feed the make activity and, more generally, to provide physical resources to extend the capacity of activities or to replace some of the physical resources. In this section, we consider that the responsibility of the buy activity is only to provide raw material to m manufacturing units, and that the raw material can by bought from r different suppliers. Indeed, a preliminary task consists of selecting the suppliers. This selection is based on various criteria, such as

- The average prices of the raw material provided
- The quality of the raw material delivered by the supplier since it influences the quality of the finished product
- The willingness of the supplier to closely participate in the activities of the supply chain
- The technical competency of the provider (are its employees skilled?)
- The resources available in the suppliers, companies
- The flexibility of the supplier
- The lead time of each supplier
- The maximal and minimal quantities that can be delivered by each supplier

3.3. An Example of Mathematical Formulation

When the selection of the e suppliers is completed, the following data are available (remember that we assume that only one new type of product is under consideration):

- $m_v, v = 1, 2, \ldots, e$ is the minimal quantity of additional raw material the supplier v is prepared to deliver during each elementary period.
- $M_v, v = 1, 2, \ldots, e$ is the maximal quantity of additional raw material the supplier v can provide during each elementary period.
- The quantity $\Omega_k, k = 1, 2, \ldots, m$ to be manufactured by manufacturing unit k during one elementary period. These variables were defined for the make activity.
- The cost $a_{vk}(x), v = 1, 2, \ldots, e; k = 1, 2, \ldots, m$ for delivering a quantity x of raw material to manufacturing unit k if provided by supplier v:

$$a_{vk}(x) = \begin{cases} 0 & \text{if } x = 0 \\ b_{vk} + c_{vk}(x) & \text{if } x > 0 \end{cases} \quad (3.9)$$

where b_{vk} is a real positive value, and $c_{vk}(x)$ is an increasing, positive, and concave function for $x > 0$. $c_v, v = 1, 2, \ldots, e$ is the cost required to increase the maximal capacity of provider v by one.

If $\sum_{k=1}^{m} \Omega_k \leq \sum_{v=1}^{e} M_v$, the capacity of the existing resources is enough to absorb the new project; otherwise, it is necessary to increase the capacity of one or more suppliers.

Assume that the capacity of the existing resources is enough to absorb the new project, i.e., to feed each manufacturing unit k with the raw material required for manufacturing Ω_k units of the new product. In this case, by solving the following problem one can select the providers to use among those that were initially selected:

$$\text{Min} \left(\sum_{v=1}^{e} \sum_{k=1}^{m} [a_{vk}(x_{vk})] \right) \quad (3.10)$$

s.t.

$$\sum_{v=1}^{e} x_{vk} = \Omega_k \quad \text{for } k = 1, 2, \ldots, m \quad (3.11)$$

$$m_v \leq \sum_{k=1}^{m} x_{vk} \leq M_v \quad \text{or} \quad \sum_{k=1}^{m} x_{vk} = 0 \quad \text{for } v = 1, 2, \ldots, e \quad (3.12)$$

Criterion (3.10) expresses the fact that the total cost should by minimized. Constraint (3.11) indicates that each manufacturing unit k must receive the quantity of raw material required for manufacturing Ω_k units of the product per elementary period. Finally, constraint (3.12) expresses the fact that a supplier either is not selected or delivers a quantity of raw material that lies between two bounds: a lower bound, which is the minimal quantity the supplier is prepared to sell during each elementary period, and an upper bound, which is the maximal quantity the supplier can deliver during one elementary period.

Because cost functions are concave, we can prove that the variable values are either maximal or minimal, except maybe for some of the values that complete the consumption of the manufacturing units.

Now, assume that the capacity of the existing resources is insufficient to perform the new project. In this case, we have to buy new resources in order to increase the capacity of some of the suppliers. The missing capacity is $\mathrm{MC} = \sum_{k=1}^{m} \Omega_k - \sum_{v=1}^{e} M_v$. This missing capacity should be added for some of the suppliers. We denote by w_v, $v = 1, 2, \ldots, e$ the increase in capacity assigned to supplier v, and T denotes the number of elementary periods during which the financial evaluation is performed. Taking into account the interest rate α, finding the minimal cost related to the buy activity consists of minimizing

$$\mathrm{CB}(w_v, x_{v,k}, v = 1, \ldots, e; k = 1, \ldots, m)$$

$$= (1+\alpha)^T \sum_{v=1}^{e} w_v \cdot c_v + \frac{(1+\alpha)^{T+1} - (1+\alpha)}{\alpha} \sum_{v=1}^{e} \sum_{k=1}^{m} [a_{vk}(x_{vk})] \tag{F5}$$

under constraint (3.11) and

$$m_v \leq \sum_{k=1}^{m} x_{vk} \leq M_v + w_v, \quad v = 1, 2, \ldots, e \tag{3.13}$$

$$\sum_{v=1}^{e} w_v = \mathrm{MC} \tag{3.14}$$

Constraint (3.13) expresses the fact that the quantity bought from each supplier should not exceed its capacity (including the possible extension of this capacity). Constraint (3.14) expresses the fact that the total of the additional capacities assigned to the suppliers is equal to the missing capacity.

In this section, we established the relationship between the quantities to be manufactured during each elementary period and the quantities of raw material to be purchase from providers. The criterion concerns horizon T.

It includes the operational cost of the buy activity and the purchasing cost of additional capacity.

3.3.2. MANAGEMENT BY DEPARTMENTS VERSUS SUPPLY CHAIN APPROACH

We first present the logical chart that shows how information is processed during the five activities. We then explain how an organization, by department, would process the information. Finally, we explain how an ideal supply chain management process would perform the same decision making.

3.3.2.1. Information Processing Constraints

In the simple model under consideration, the sell activity determines the quantities of product to be sold among the retailers of the region. Thus, the sell activity derives the quantities $\bar{d}_{ij}$ that should be carried to each retailer j of each region i during each elementary period from the random demand d_i in region i.

In our example, the move activity is limited to the transportation of WIP inside the manufacturing units and to the transportation from manufacturing units to retailers. The transportation from suppliers to manufacturing units is reduced to a cost that is integrated in the buy activity. The move activity determines the flows q_{kij} from manufacturing units to retailers, taking into account the average quantities $\bar{d}_{ij}$ sold by retailers and the quantities Ω_k produced by the manufacturing units during each elementary period.

The store activity concerns only the inventories located at the retailers. It consists of keeping the safety stock at each retailer to absorb the randomness of the demand. The safety stock y_{ij} to be kept at retailer j of region i is derived from the random demand d_{ij} and the constant replenishing of finished products $\bar{d}_{ij}$.

The make activity concerns manufacturing products. At the strategic level, which is the level of interest in this chapter, the objective is to define the quantities Ω_k of products to be manufactured by manufacturing unit k during each elementary period. These quantities are defined taking into account the random quantities d_i required in each region i during each elementary period.

The buy activity consists of selecting the suppliers v and defining the quantities w_k to be bought from each of them during each elementary period, taking into account the quantities Ω_k produced by the manufacturing units. Indeed, costs play a part in each decision made at the strategic level.

In the sell activity, our model integrates the profit, which is the difference between the revenue of this activity and the expenses made to run this activity.

The move activity has several types of expenses, namely,

- The expenses for buying new carts to carry WIP inside the manufacturing units
- The expenses incurred while performing the move activity inside the manufacturing units
- The expenses for buying new trucks to carry finished products from manufacturing units to retailers
- The expenses incurred while performing the move activity between manufacturing units and retailers

The strategic decisions made during the store activity have to take into account the costs for keeping in stock one unit of product during one elementary period at each retailer.

The make activity has two types of expenses: expenses to increase the capacity of some of the manufacturing units and expenses to run the manufacturing units. Finally, the purchasing costs are integrated into the decisions concerning the buy activity.

Figure 3.4 shows the logical connections between the information processing stages at the strategic level. Costs and information not used in another model are not indicated. In Section 3.3.2.2, we show how a system organized into departments processes the information:

- To assign a production level to each manufacturing unit
- To select the suppliers for each manufacturing unit
- To select the retailers in each region
- To define the transportation facilities and their use
- To define the safety stocks

In Section 3.3.2.3, we show how the same problem can be solved in a supply chain environment.

3.3.2.2. Information Processing If the System is Organized in Departments

Considering the flow chart in Fig. 3.4, it may seem that two activities can be optimized first — namely, the make and sell activities — since they require only the d_i values and the costs. Thus, the objective for our example and assuming that each activity is a department, is to minimize cost (Eq. F4), taking into account constraints (3.7) and (3.8) and the costs associated with the model (make activity).

3.3. AN EXAMPLE OF MATHEMATICAL FORMULATION

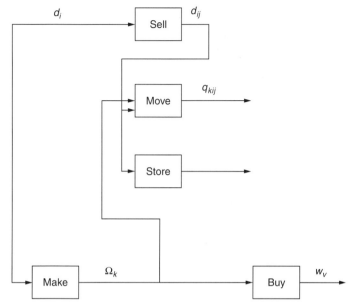

FIGURE 3.4 Data processing constraints

Optimizing the make activity leads to

- The production level of each manufacturing unit (i.e., the Ω_k values).
- Determination of the number of new resources to buy in each manufacturing unit. We denote the optimal cost by OP(make).
- Maximization of local profit (or minimization of local loss) (Eq. F1) taking into account constraints (3.1)–(3.3) and the costs associated with the model (sell activity).

By optimizing the sell activity, the quantity to be sold by each retailer j of each region i can be determin. The optimal value of the profit (or loss) for this activity is denoted by OP(sell).

These two optimization problems lead to quantities Ω_k and d_{ij}, respectively. At this point of the computation, it is possible to optimize the three other models.

The model corresponding to the buy activity is optimized taking into account quantities Ω_k. It consists of minimizing cost (Eq. F5) taking into account Eqs. (3.9)–(3.14). This optimization leads to the selection of suppliers and the definition of the quantities of raw material that will be provided by each of them during each elementary period (i.e., quantities x_{vk}). The minimal value of Eq. (F5) is denoted by OP(buy/make). This notation reflects the fact that the computation of the optimum cost is based on the results of the optimization of the make activity: The optimum is conditional on the decisions made in the make activity.

The model corresponding to the move activity is optimized in parallel with the previous one since the quantities Ω_k and d_{ij} are known at this stage of the computation. Optimizing this model consists of minimizing cost (Eq. F2) taking into account constraints (3.4) and (3.5) as well as the purchasing and running costs related to this activity. This optimization determines

- The number of vehicles to purchase for each manufacturing unit in order to deal with the additional load of WIP
- The number of trucks to purchase for transporting finished products from manufacturing units to retailers
- The flows of finished products from manufacturing units to retailers (i.e., the q_{kij} values)

The minimal value of Eq. (F2) is denoted by OP(move/[make and sell]). This means that the optimization of the move activity depends on the decisions made for the make and sell activities.

The model corresponding to the store activity requires the d_{ij} values to be optimized. Optimizing this model consists of minimizing cost (Eq. F3) under constraint (3.6). From this optimization, the possible additional storage capacities for retailers and the additional target inventories of each of the retailers can be determined.

The minimal value of Eq. (F3) is denoted by OP(store/sell), which means that the optimal decisions made in the store activity depend on the decisions previously made in the sell activity.

Finally, the decisions made at the strategic level for the different activities lead to a total profit of

$$P(DEP) = OP(sell) - OP(make) - OP(buy/make)$$
$$- OP(move/[make \text{ and } sell]) - OP(store/sell) \qquad (F^*)$$

This profit is usually not optimal for the whole system since the last three elements of the sum of the right side of Eq. (F*) are constrained by the results of the optimization of the make and sell activities.

3.3.2.3. Information Processing in the Supply Chain Environment

In a supply chain environment, the preliminary analysis in each activity is the same as in an organization by departments, but the mathematical model is unique: It is the concatenation of the five models of the activities.

3.3. AN EXAMPLE OF MATHEMATICAL FORMULATION

As a result, the problem to solve is the maximization of the sum of profits (minimization of costs) in Eqs. (F1)–(F5) under constraints (3.1)–(3.14). Thus, the total profit of the system is obtained by maximizing the following function simultaneously under constraints (3.1)–(3.14):

$$P(\text{supply chain}) = \max(BN - CM - CS - MAN - CB) \qquad (F^{**})$$

where BN is derived from Eq. (F1), CM from Eq. (F2), CS from Eq. (F3), MAN from Eq. (F4), and CB from Eq. (F5).

It can easily be seen that the total profit will be higher in Eq. (F^{**}) than in Eq. (F^*). If we fix the production levels at the different plants and the demand that each retailer will fulfill (Ω_k and d_{ij}, respectively) without considering how this production will be allocated between the manufacturing units and the retailers (i.e., the quantities q_{kij}) and how much inventory will be carried by each retailer (y_{ij}), the results can be suboptimal.

It is possible that profit margins for a retailer are high. A departmental approach tends to maximize the allocation of supply to this retailer. However, if the demand fluctuations are also high for this retailer, then to meet a high volume of demand with sufficiently high customer service levels this retailer may have to carry very high inventories. If the inventory carrying costs at this retailer are high, then this decision can be very costly for the supply chain as a whole.

Similarly, consider a manufacturing unit that has a low manufacturing cost and a high resource capacity. A departmental approach to decision making will tend to run this manufacturing unit at its maximum capacity. However, consider the fact that this manufacturing unit is located away from the markets whose demand it is going to meet. In this case, the transportation costs from this plant to the retailers may be prohibitively expensive.

These are just two examples, but they are sufficient to show that a supply chain approach to strategic decision making is much more efficient than a departmental approach.

3.3.2.4. Sensitivity Analysis: A Method to Adjust the Control Parameters of a Supply Chain

The complexity of a supply chain control problem is much greater than the complexity of the five subproblems to be solved in an organization with departments. However, the availability of sophisticated software solutions and powerful computers coupled with recent research in this domain can make this seemingly daunting task feasible to obtain approximately optimal solutions, if not absolutely optimal ones. Some of the approaches are based on decomposition, and others are based on approximations or simplification. The approach by departments presented previously is a type of decomposition approach and this approach can lead to a solution that is not optimal. Simplification approaches consist of working on models that are easy to handle and that are expected to lead to a "good" solution despite the fact that they may not represent the real system. Simplification approaches consist of replacing the criteria or constraints that are difficult to handle by criteria or constraints that fit with well-known algorithms.

When random events are included in the process under consideration, as is the case for the model illustrated in this section since the demands on the retailers are random, sensitivity analysis or simulated annealing usually lead to near-optimal results. Here, we provide a brief insight into sensitivity analysis.

Consider a real vector $X = \{x_1, x_2, \ldots, x_h\}$ and a process f that transforms X into a value denoted by $f(X, \varepsilon)$, where ε is a random variable. The process can be a function of any type that includes random variables or an algorithm that includes some random events. Assume that the goal is to find X^* such that

$$\overline{f(X^*, \varepsilon)} = \min_{X \in D} \overline{f(X, \varepsilon)}$$

where D is the domain of definition of X, $\overline{f(X, \varepsilon)}$ is the mathematical expectation of $f(X, \varepsilon)$, and X is a feasible solution. Let $W_i = \{0, 0, \ldots, \omega, 0, \ldots, 0\}$ (ω is the ith element) be a vector of dimension h, where ω is a real value chosen at random. Then compute $f(X, \varepsilon)$, T times. We obtain $f_1, f_2, \ldots, f_T$. If T is large enough, we can write $\overline{f(X, \varepsilon)} \approx 1/T(\sum_{k=1}^{T} f_k)$. For $i = 1, 2, \ldots, h$, we compute $f(X + W_i, \varepsilon)$, T times where $X + W_i \in D$, which leads to $f_1^i, f_2^i, \ldots, f_T^i$. We derive an evaluation of the mathematical expectation $\overline{f(X + W_i)} \approx 1/T(\sum_{k=1}^{T} f_k^i)$ for $i = 1, 2, \ldots, h$. If $X + W_i \notin D$, we set $\overline{f(X + W_i)} = +\infty$.

Let $\overline{f(X + W_{i*})} = \min_{i \in \{1,2,\ldots,h\}} \overline{f(X + W_i)}$. If $\overline{f(X + W_{i*})} < \overline{f(X)}$, then we set $X = X + W_{i*}$ and we restart the process; otherwise, we keep X as the solution and $\overline{f(X, \varepsilon)}$ the value of the criterion.

The following algorithm summarizes the previous discussion:

1. Choose, at random, a feasible solution X to the problem.
2. Choose ω. The choice of ω depends on the problem. It should be small enough to fit the function that is "optimized."
3. Simulate the process $f(X, \varepsilon)T$ times to obtain $f_1, f_2, \ldots, f_T$ and simulate $f(X + W_i, \varepsilon)T$ times to obtain $f_1^i, f_2^i, \ldots, f_T^i, i = 1, 2, \ldots, h$.

4. For each $t \in \{1, 2, \ldots, T\}$, we compute $f_t^1, f_t^2, \ldots, f_t^h, f_t$ in the same run to use the same sequence of random events.
5. Compute $\overline{f(X, \varepsilon)} \approx 1/T(\sum_{k=1}^{T} f_k)$ and
   $\overline{f(X + W_i)} \approx 1/T(\sum_{k=1}^{T} f_k^i)$.
6. Compute $\overline{f(X + W_{i*})} = \min_{i \in \{1, 2, \ldots, h\}} \overline{f(X + W_i)}$.
   If $\overline{f(X + W_{i*})} < \overline{f(X)}$, then
     6.1. Set $X = X + W_{i*}$.
     6.2. Go to step 3.
   Otherwise,
     6.3. X is the solution to the problem and $\overline{f(X, \varepsilon)}$ is the value of the criterion.
     6.4. End of the program.

3.3.2.5. Application of Sensitivity Analysis to Our Specific Problem

In the problem illustrated in Section 3.3, the demand d_i, $i = 1, 2, \ldots, r$ in each region i is a random variable that corresponds to the randomness represented by ε in Section 3.3.2.4. This random variable is known. We first

- Generate a feasible solution to the sell problem
- Generate a feasible solution to the make problem

The feasible solutions are generated at random. Knowing the previous feasible solutions, we then

- Compute the optimal solution to the move problem
- Compute the optimal solution to the store problem
- Compute the optimal solution to the buy problem

At this point of the computation, it is possible to compute the total cost associated with the previous solution. Let X be the initial feasible solution and f_1 be the total cost.

The next step of the process consists of computing $X_j = X + W_j$, where $W_j = \{0, 0, \ldots, \omega, \ldots, 0\}$ (ω is the jth elements) and ω is generated at random in $\{-1, 1\}$, for $j = 1, 2, \ldots, r$. We apply to each X_j, $j = 1, 2, \ldots, r$, the computation we previously applied to X, and we obtain the costs that we denote by $f_1^1, f_2^2, \ldots, f_1^r$, respectively. These values can be evaluated in the same run as f_1.

Starting from a new generation of a set $\bar{d}_i$, $i = 1, 2, \ldots, r$, and by making the same sequence of computation, we obtain a new set $f_2, f_2^1, f_2^2, \ldots, f_2^r$, and so on.

The example presented in Section 3.3 consists of analyzing the introduction of a new type of product in a supply chain and illustrating the difference between an approach by departments and an approach in a supply chain environment.

In each of the five activities, a preliminary study leads to a selection of resources that may be required for implementing the new project and to the definition of the incremental costs that must be taken into account in the model.

The approach by departments consists of solving one optimization problem for each of the five activities in an order that is partially set by the order in which data are generated. This approach is biased since some of the optimization problems are constrained by the result of the optimization of other problems.

The approach developed in the supply chain environment is global and thus more complex, but the complexity is limited by the fact that the existing state of the system is not reconsidered: The approach is incremental. The total cost resulting from this approach is obviously less than that of the previous approach.

We did not propose methods that could be used to optimize the five local problems (approach by departments) or the global problem (supply chain environment): It is not the goal of this section, and numerous Operations Research (OR) techniques or heuristic approaches are available in the literature.

Indeed, the introduction of a new project concerns the strategic level, but the same difference in the approaches can be shown at the tactical level, which is presented in Chapter 4. In fact, the models are similar at the strategic level and the tactical level. The main differences are that there are no investments at the tactical level, and the horizon at the tactical level is shorter than at the strategic level.

Models are much more detailed at the tactical level than at the strategic level. An example of tactical model for scheduling is presented in the Appendix.

3.4. DOMINANT PARTNER

Real life is quite different from the ideal situation. When the activities are not performed by the same company, competition arises, and usually the strongest partner becomes dominant and imposes its strategy on the others. It is this tendency that we address here.

3.4.1. WHO IS THE DOMINANT PARTNER?

Most supply chains have a dominant partner that has a significant impact on the behavior of the supply chain. The dominant partner can be the raw material supplier, the manufacturer, the transporter, the wholesaler, or the retailer (corresponding to the five functions of buy, make, move, store, and sell). However, it has been observed that in most cases either the manufacturer or the retailer is the dominant player.

Good examples of supply chains in which the retailer is the dominant player are the supermarkets such as Wal-Mart and K-Mart in United States and Tesco in Europe. In most of these cases, the products involved are consumer goods, which are typically high-volume, standardized goods that do not have very complex and lengthy manufacturing processes and have small manufacturing lead times. In most cases, the manufacturing is make-to-stock.

Good examples of supply chains in which manufacturers are the dominant players are the heavy equipment manufacturers. The products involved in these types of supply chains have very complex and lengthy manufacturing processes and long manufacturing lead times. The manufacturing operations tend to be make-to-order, although large-volume products may still follow the make-to-stock or assemble-to-order policy. Examples include companies such as heavy machinery manufacturer Cincinnati Milcron, construction and heavy equipment manufacturer Caterpillar, and farm equipment manufacturer John Deere.

In between these two extremes are the supply chains in which manufacturers and retailers have a comparable say. The products in these supply chains are moderately complex; however, the manufacturers do not have their own retail outlets. The manufacturing process for these products is complex, however, most of the end products are assembled from common components. These products lean toward an assemble-to-order type of environment. Examples of these types of supply chains include computer and electronic goods manufacturers and the electronic goods retailers. A good example of this type of supply chain in the United States is the relationship between computer manufacturer Compaq and electronic goods retailers Circuit City and Best Buy. Establishing the dominant partner in this case is not straightforward.

3.4.2. WHY IS A PARTNER DOMINANT?

A closer examination of these scenarios reveals that the dominant player in the supply chain is the one that is the closest to the consumer. The closeness here is not in terms of physical proximity but in terms of hearing the customers' voice and responding to it. As emphasized throughout this book, the ultimate goal of the supply chain is customer satisfaction. The partner in the supply chain that can understand the consumer demand and fulfill it in a timely and cost-effective manner is the dominant player in the supply chain. In other words, the partner that controls consumer information and can deliver in response to it dominates the supply chain. There are four main components of this consumer information:

- What do they want?
- When do they want it?
- How much do they want?
- How much are they willing to pay for it?

What do they want? This is probably the most difficult of the four questions. It has two subquestions: What are the customer needs? and What product/product features would satisfy those needs? Adding to the complexity is the fact that different people want different product features to satisfy the same needs.

When and how much do they want? The other important piece of information is when would the customers need the product and in what quantities. Would the customers be willing to wait for the product if it is not available (or not in the amount/quantity they want)? For how long would they be willing to wait?

How much are they willing to pay for it? The last piece of the puzzle is the price they are willing to pay for the product. Would they go to the competitors if they have a lower price? How much lower could the competitor price be to attract the customers to switch?

3.4.3. DOMINANT PARTNERS AND PRODUCTION TYPES

In light of the previous discussion, we revisit the three types of supply chains discussed in the previous section. The retailers dominate the supply chains for consumer goods types of products. The answer to the question what do the customers want is well-known. There is a range of products with different features to satisfy the needs of the different types of customers. Typically, there are many manufacturers in the market that can supply comparable products. Thus, the manufacturers have little control over this information and even less control over the supply (the competitors are always willing to step in). The main control the manufacturers have in this type of market is dominance of their brand names. Because different manufacturers provide products with similar features and quality, the products are viewed by customers as commodities. Customers typically do not want to wait for the product to become available and will easily substitute a product with a product from a competitor that offers a lower price for the same product features and quality. Thus, it is important that the products be available at the time, in the quantity, and at the price the customer wants. Since retailers have the information about these three requirements and, more important, they can deliver them to the customer, they dominate these supply chains.

The manufacturer dominates the supply chain in make-to-order type of products. In these cases, the most important piece of information is what customers want. Sometimes, even the customers are not sure what products can satisfy their needs. In these cases, manufacturers having a strong knowledge of the technology and the capability to deliver have the upper hand. They interact directly with customers to understand their needs and offer customized solutions/products. The customer is willing to wait and pay a premium for quality, reliability, and customized products. Often, the manufacturers sell directly to the customers.

In the third case, in which the manufacturer and the retailers have a comparable say, both hold pieces of the information puzzle. The manufacturing of most of the products involved in these supply chains requires complex and often proprietary technology. There are only a few manufacturers and they hold the key to the question of what customers want, in terms of both understanding the needs of the customers and having the technology to satisfy these needs with their products. However, the customers do not want to wait too long for the products and, if possible, would like to buy them off the shelf. The retailers play an important role in satisfying customers needs at the time they want it and in the quantities they want. Thus, the retailers hold the second half of the information puzzle and, hence, have a comparable say in the supply chain.

To achieve the real benefits from a supply chain, the constituents with the potential to dominate should resist the urge to do so. They should instead try to share the information and other advantages they have with other constituents to create a level of comfort for other participants. Only under the umbrella of a sharing process, as discussed in Section 3.2, would the other constituents of the supply chain be encouraged to participate more wholeheartedly. This would result in an overall increase in the efficiency and competitiveness of the supply chain.

> Dominant partners are the result of competition when different companies perform supply chain activities. The partner that will dominate depends on the type of production. The retailers dominate the supply chains for consumer goods types of products. The manufacturers dominate the supply chain for make-to-order type of products.
>
> The explanation is simple. The dominant partner is the one that can understand the consumer demand and fulfill it in a timely and cost-effective manner. Thus, the dominant partner is the one that is the closest to the consumer, and this depends on the type of production considered.

3.5. CONCLUSION

In this chapter, we discussed the high-level decisions that are needed in the five main activities of the supply chain (buy, make, move, store, and sell) at the strategic level to make a successful supply chain. We presented a high-level supply chain architecture and the steps to attain this architecture. We also discussed why, in the real world, it is difficult to have a balanced supply chain because of the presence of a dominant partner. This tendency to dominate, however, should be avoided in favor of the sharing process.

REFERENCES

Ball, M. O., and Taverna, R., Sensitivity analysis for the matching problem and its use in solving matching problems with a single side constraint (transit crew scheduling), *Ann. Operations Res.* **4**, 25–56, 1985.

Carter, J. R., and Narasimhan, R., "A comparison of North American and European future purchasing trends," *Int. J. Pruchasing Materials Management* **32**, 12–22, 1996.

Choi, T. Y., and Hartley, J. L., "An exploration of supplier selection practices across the supply chain," *J. Operations Management* **14**, 333–343, 1996.

Cohen, M. A., and Mallik, S., "Global supply chains: Research and application," *Production Operations Management* **6**, 193–210, 1997.

Ellram, L. M., and Cooper, M. C., "Supply chain management, partnership, and the shipper–third party relationship," *Int. J. Physical Distribution Logistics Management* **1**(2), 1–10, 1990.

Fisher, M. L., "What is the right supply chain for your product?" *Harvard Business Rev.* **75**, 105–116, 1997.

Gattorna, J. (Ed.), *Strategic Supply Chain Alignment. Best Practice in Supply Chain Management.* Gover, Aldershot, UK, 1998. [ISBN 0-566-07825-2]

Gentry, J. J., "The role of carriers in buyer–supplier strategic partnerships: A supply chain management approach," *J. Business Logistics* **17**, 33–55, 1996.

References

Glazebrook, K. D., "Sensitivity analysis for stochastic scheduling problems," *Mathematics Operations Res.* **12**(2), 205–223, (1987).

Holmlund, M., and Kock, S., "Byers-dominated relationships in a supply chain—A case study of four small-sized supplier," *Int. Small Business J.* **15**, 26–40, 1996.

Klemt, W. D., "Schedule synchronization for public transit networks," in *Proceedings of the Fourth International Workshop on Computer-Aided Scheduling of Public Transport*, Springer-Verlag, pp. 327–335, 1988.

Knoff, R., and Neuman, J., "Quick response technology: The key to outstanding growth," *J. Business Strategy*, 61–64, September/October, 1993.

Kolen, A. W. J., and Rinnooy Kan, A. H. G., Van Hoesel, C. P. M., and Wagelmans, A. P. M., "Sensitivity analysis of list scheduling heuristics," *Discrete Applied Mathematics*, **55**(2), 145–162, 1994.

Kuglin, F. A., *Customer Centered Supply Chain Management: A Link by Link Guide*. American Management Association, New York, 1998.

Lalonde, B., "Small shipments—manage the supply chain," *Transportation Distribution* **37**, 11, 1996.

Leenders, M. R., Nollet, J., and Ellram, L. M., "Adapting purchasing to supply chain management," *Int. J. Physical Distribution Logistics Management* **24**, 40–42, 1994.

Levy, D., "Lean production in an international supply chain," *Sloan Management Rev.* **38**, 94–102, 1997.

Lummus, R. L., and Alber, K. L., "Supply chain management: Balancing the supply chain with customer demand," *APICS* Series on Resource Management, pp. 3–5. St. Lucie Press, Boca Raton, FL, 1997.

Mentzer, J. T., "Managing channel relations in the 21st century," *J. Business Logistics* **14**(1), 31, 1993.

Poirier, C. C., and Reiter, S. E., *Supply Chain Optimization. Buliding the Strongest Total Business Network*. Berret-Koehler, San Francisco, 1996.

Porter, M., *Competitive Advantage: Creating and Sustaining Superior Performance*. Free Press, Macmillan, 1985.

Preiss, K., Goldman, S. L., and Nagel, R. N., *Cooperate to Compete: Building Agile Business Relationships*. Van Nostrand–Reinhold, New York, 1996.

Riggs, D. A., and Robbins, S. L., *The Executive's Guide to Supply Chain Management Strategies: Building Supply Chain Thinking into All Business Processes*. American Management Association, New York, 1998.

Scott, C., and Westbrook, R., "New strategic tools for supply chain management," *Int. J. Physical Distribution Logistics Management* **21**, 22–33, 1991.

Tagaras, G., and Lee, H. L., "Economic models for vendor evaluation with quality cost analysis," *Management Sci.* **42**, 1531–1543, 1996.

Towill, D. R., "The seamless supply chain—The predators strategic advantage," *Int. J. Technol. Management* **13**, 37–56, 1997.

Treacy, M., and Wiersema, F., *The Discipline of Market Leaders*. Addison-Wesley, Reading, MA, 1995.

Walton. L. W., "Partnership satisfaction: Using the underlying dimensions of supply chain partnership to measure current and expected levels of satisfaction," *J. Business Logistics* **17**, 57–75, 1996.

4

SUPPLY CHAIN AT THE TACTICAL LEVEL

4.1. INTRODUCTION

This chapter is devoted to the tactical issues in a supply chain. This is the level at which medium-term decisions are made regarding the way different activities will be performed and organized. This chapter covers the actions to be taken in order to obtain a supply chain that fits with the definition given at the beginning of Chapter 2 — that is, an organization that encapsulates all the partners required to transform raw material into products-in-use in an efficient manner. This implies that information and material flow freely at low cost and high speed. Such a system should include suppliers and customers, and all the partners are supposed to cooperate in order to maximize the efficiency of the whole system, despite the fact that some of them may compete.

In Section 4.2, we focus on how partners of a supply chain influence each other through their decisions. The fact that partners strongly influence each other calls for fair cooperation, which in turn requires some agreements between the partners (this aspect was discussed in Chapter 3). In Section 4.3, we analyze the actions needed at the tactical level to achieve a successful supply chain. The goal of these actions is the implementation of the mechanisms required to meet the supply chain definition. In Section 4.4, we investigate the criteria that should be considered when evaluating a supply chain. Costs and benefits are important in the short term, but other criteria that guarantee long-term success cannot be ignored. Finally, the method to evaluate profits is developed in Section 4.5.

4.2. LOCAL DECISIONS AND GLOBAL CONSEQUENCES

Cooperation is the key word for the success of a supply chain. To cooperate fruitfully, a partner should first realize that any decision made in one of its departments influences not only the company as a whole but also other partners in the chain. The following examples illustrate this fact:

1. Setting a limit on inventory levels in a company with a seasonal demand pattern may result in lower inventory costs, but the additional production cost incurred when demand is at a higher level may more than offset the inventory cost saving. Furthermore, if outsourcing is used to meet the shortfall in production, it may introduce delays or quality problems and thus disturb the organization of the partners of the supply chain.

2. Firing skilled employees during periods of low demand and production will result in salary savings. However, the additional cost incurred by using subcontractors or hiring skilled employees at a higher salary when the activity restarts may more than offset salary savings. It may also disturb the organization of the whole supply chain due to the delays that may occur in the production.

3. Buying raw materials at a low price when demand is low may result in excessive inventory costs since the company must store the material until demand increases. Thus, production costs may increase, reducing the efficiency of the whole supply chain.

4. Consider the activity (performed by a partner or a department of a partner company) concerned with manufacturing of products. The decisions that can be taken about this activity are mainly a change in the production mix or a change in the production levels. These decisions are constrained not only by the capacity of the system (usually the bottleneck resources and the work force) but also by (i) the raw material availability, which depends on those in charge of buying raw materials or components from outside the supply chain; (ii) the possibility of selling (or not selling) the additional production quantities, which may call for new advertising, new skills, and new maintenance management, and thus call on the competencies of other activities of the supply chain; (iii) the possibility of introducing new storage facilities and/or new storage management: this depends on the partner in charge of inventory; (iv) the transportation capacity, which concerns the logistics activity.

5. The objective of the partner involved with the selling activity is to increase the sales of the product. Some of the techniques used for increasing sales are promotions and reduced price sales. These activities can increase the demand for the product over a short period of time. However, if the manufacturing activity and the storing activity have not been informed about the expected surge in demand, they may not be able to meet that demand. This means customer demand would go unfulfilled and many customers would be disappointed. They may then buy products from competitors.

These examples show that a decision resulting in short-term benefit for one partner could penalize other partners in the short term or the same partner in the medium term.

Note that a decision made concerning the manufacturing activity influences the state of the other activities — that is, their constraints and/or their flow of material and information. This calls for a decision in each of these activities to adjust their behavior to the new situation. In turn, this modifies the constraints and the flows of material and information applied to all the activities, including the manufacturing activity. This is true for any activity (performed by a partner or a department of a partner company) of the supply chain. This iterative aspect has to be analyzed to understand the behavior of the whole system.

Figure 4.1 represents this situation. Partners are represented by A_0, $A_1, \ldots, A_k$. We assume that partner A_0 makes a decision. This decision applies to system A_0 but also disseminates among all the partners' systems, altering more or less the states of these systems (i.e., their material or information flows as well as their constraints). These first alterations call for a first set of decisions denoted by $D(0, 1), D(1, 1), \ldots, D(k, 1)$ that apply to $A_0, A_1, \ldots, A_k$, respectively. The goal of these decisions is to adjust the behavior of the systems to the new situation. In turn, these decisions disseminate, altering all the partners' systems, calling for a new set of decisions $D(0, 2), D(1, 2), \ldots, D(k, 2)$, and so on. The effects of the initial decision usually dampen quickly as time progresses. Indeed, other decisions made by the same or other partners also influence the whole system, and their effects are combined with those of other decisions. This shows the complexity of the management of a supply chain.

When two partners of a supply chain share the same market segment, they influence each other not only in terms of organization but also in terms of production level and, thus, direct incomes. A solution could be either to have all-out competition between companies or to control the system, for instance, by sharing the market among the partners. In our opinion, none of these diametrically opposed solutions lead to a successful supply chain.

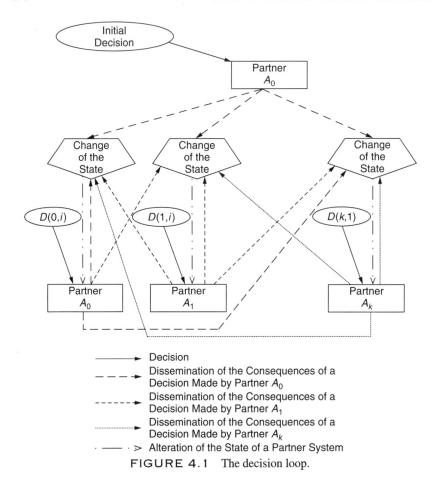

FIGURE 4.1   The decision loop.

Consider the first solution. If the system is of the master–slave type, as is the Benetton organization, then the leader uses the competition among the partners for its own profit or else competitors spend much energy fighting with each other to gain leadership. This solution is not optimal in the long term when applied to technology-sensitive systems as it does not favor the cooperation of all the partners for technical innovation.

Too much competition may lead the partners to compete on cost in the short term and cut down expenditure on research and development activities.

The second solution, which consists of planning the market for the supply chain partners, is obviously disastrous since it destroys competition and thus kills innovation. This has been proven time and again for about a century in different countries with controlled economies. The best solution is certainly somewhere in between.

In any case, whether the supply chain partners are competitors or not, cooperative arrangements that tie firms to each other and that tie their success or failure to the success or failure of the whole supply chain should be incorporated at the strategic level. The goal is to guarantee fair cooperation between the partners. This was discussed in Chapter 3.

> A decision made by any partner of the supply chain disseminates among the whole supply chain. This means that such a decision requires adjustment decisions from other partners. As a consequence, a global information system is necessary to allow all the partners to be informed in real time of the state of the system and decisions made anywhere in the system. Also, each partner should accept the rules derived from the cooperation arrangements decided at the strategic level. The goal of these rules is to ensure that each one of the partners is prepared to adjust to any decision that complies with these rules.

4.3. BUILDING AN EFFICIENT SUPPLY CHAIN AT THE TACTICAL LEVEL

4.3.1. TACTICAL AND STRATEGIC LEVELS

We assume that a sharing mechanism exists among the partners, and that the supply chain encapsulates all the partners that are involved in the production chain. The goal of this section is to highlight the conditions for information and material to flow freely and to adequately coordinate the activities of the partners. Thus, this section concerns the tactical activities (also called secondary activities) as opposed to the strategic activities (also called primary activities) that were discussed in Chapter 3.

Strategic activities include research, technology development, human resources strategy, physical resource procurement, and marketing plan. These activities require a close collaboration of the partners of the supply chain and apply to all the components of the supply chain. They are developed in the framework resulting from the agreements established between partners, and are the guiding factor for tactical activities.

Tactical activities are those that are encountered in production management, with this term being used in its broader sense. These include manufacturing and logistics activities, applied marketing, day-to-day human resources management, sales, quality management, and inventory management.

Tactical activities are further grouped into the following macroactivities:

- Buying, which involves raw material and component procurement as well as any activity that is involved with acquiring resources for the supply chain
- Making, which groups manufacturing and related activities, such as quality control and maintenance
- Moving, which involves the transportation of goods and personnel inside and outside the supply chain
- Storing, which involves storing products at all of its transformation levels, from raw material to finished products
- Selling, which includes commercial activities, marketing, and more generally all activities that concern and/or influence the outputs of the supply chain

Both strategic and tactical activities are horizontal in the sense that each one concerns directly or indirectly each of the partners of the supply chain, but strategic activities also cut across the tactical activities and integrate them. Consider, for instance, the marketing strategy that is developed by the supply chain partner in charge of this activity, in close connection with all other partners at the highest decision level. The resulting strategic policy becomes a component of the rules that apply to each one of the partners. As a consequence, any tactical decision made by a partner should fit with this strategic policy. Assume, for instance, that the marketing strategy insists on quick responses of the supply chain to customer requirements. In this case, logistics (move activity) may be organized to work twenty-four hours a day and to use airplanes instead of trucks if necessary, manufacturing (make activity) may be based on flexibility to adapt to demand, some component or raw material inventories (store activity) will possibly be accepted to guarantee quick responses, suppliers (buy activity) will be selected with regard to their efficiency, and so on. The horizontal effect of strategic activities is clear from this example. Tactical decisions also have a horizontal effect in the sense that they disseminate among all the partners, as explained previously.

Figure 4.2 represents the effect of strategic and tactical decisions. It shows that strategic decisions "cover" all the partners and all the tactical decisions, whereas tactical decisions concern all the partners directly or indirectly.

> The decisions made at the strategic level define the set of decisions that can be made at the tactical level. Strategic decisions apply to each of the partners and are a major cohesive factor of the supply chain.

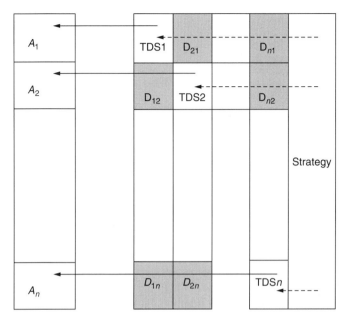

FIGURE 4.2   Decision-making system.

A_i: Activity i TDSi: Tactical Decision System for Partner i
D_{ij}: Dissemination of a Decision Made at Activity i to Activity j

4.3.2. THE TACTICAL OBJECTIVES IN A SUPPLY CHAIN

The global objective of a supply chain is customer satisfaction. At the same time, individual components of the supply chain aim to maximize their shareholder value by maximizing the return on investment (ROI) of their investors. ROI is the ratio of profit earned to capital employed during 1 year. This strategic objective can be translated into several short- and medium-term objectives at the tactical level. In other words, these tactical objectives, if reached, guarantee a large, if not maximal, ROI. The main tactical level objectives in a supply chain are

- Minimizing the time required for converting orders into cash
- Minimizing the total work-in-process (WIP) in the supply chain
- Improving pipeline visibility — that is, the visibility of each one of the activities of the supply chain by each one of the partners
- Improving visibility of demand by each one of the partners

- Improving quality
- Reducing costs
- Improving services

Some of these tactical objectives are linked to each other. Next, we examine these tactical objectives in details.

> The global objective of a supply chain is customer satisfaction while maximizing the return of investment. At the tactical level, this global objective is translated into several detailed objectives, such as minimizing the time required for converting orders into cash and minimizing WIP, improving the visibility of the pipeline and quality, and reducing costs at the various levels of the supply chain.

4.3.2.1. Minimizing the Time Required for Converting Orders into Cash

This objective is much more than reducing the production cycle — that is, the time needed to convert raw material and components into finished products. It includes the time required to obtain raw material and components, to control their quality, and to handle and, if necessary, store them until they are used. It also includes the time finished products are stored and prepared to be shipped to retailers, the transportation time, and the time they are stored again before being delivered to customers.

To fulfill an order arriving in a supply chain, it is necessary to get the raw material and the outsourced components on time and to ensure that the required resources (machines, tools, employees, transportation devices, etc.) will be available on time. Minimizing the time for fulfilling an order requires strong coordination between the selling, moving, manufacturing, storing, and buying activities.

The buying activity consists of providing raw material and outsourced components when required. The risky and expensive solution to reach this goal is to have inventories at one's disposal. This solution is risky since customers' requirements could change without notice and thus make inventories obsolete. It is expensive since inventories tie up capital and quality may suffer as a consequence of storing and handling. Thus, inventories should be avoided as much as possible in a supply chain. Selecting many reactive providers that compete with each other is the best solution to

4.3. BUILDING AN EFFICIENT SUPPLY CHAIN AT THE TACTICAL LEVEL

reduce the risk of running out of raw material or components. This solution can be improved by working with the suppliers on a "win-win" basis by providing them with incentives in return for being reactive. Examples of such incentives include the following:

- Guaranteeing a minimal level of work per year
- Offering technical support to develop complex components
- Offering training sessions
- Helping to improve quality

The logistics (move activity) as well as the store activity should cooperate with the buy activity to transport and store raw material and components.

Manufacturing a product is the responsibility of the make activity. The objective of the production management system in this activity is to ensure that the production cycle of each ordered product is as short as possible, taking into account the partial schedule that exists at the time the product is launched in production. This calls for a real-time scheduling system that takes advantage of the idle periods of the resources. In such a scheduling approach, a product is scheduled as soon as it is launched in production and the products previously scheduled are not re-examined. One such approach, which is based on recent results derived from the no-wait scheduling literature, is discussed in the appendix.

Once the manufacturing of the product is complete, it should be packaged (end of make activity), carried to the retail location (move activity), and sold (sell activity). The storage (store activity) of finished products should be reduced as much as possible since it ties up capital and requires handling that may affect the quality.

It should be noted that the goal is not only to optimize each of the previous activities but also to coordinate them. Being able to obtain raw material or components in a very short amount of time may result in unnecessary expenses if the manufacturing system is overloaded and cannot manufacture the product immediately. Similarly, if logistics resources are overloaded, it may be necessary to delay the production of the ordered product. Thus, it appears that when a product is ordered the buy, make, make and sell activities should be scheduled simultaneously, taking into account the schedule of the products that were ordered previously and the availability of external resources such as suppliers.

Indeed, this requires precise and real-time information for each of the five activities. An information system that covers the whole supply chain is thus required. The whole system described previously is represented in Fig. 4.3.

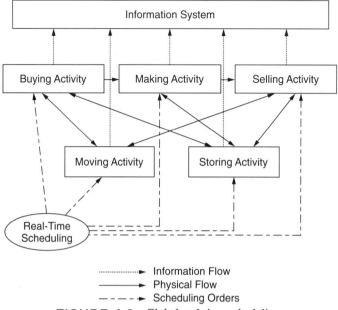
FIGURE 4.3 Global real-time scheduling.

Such a system will enable the supply chain to fulfill customer demand in the shortest possible time at the lowest possible cost by maximizing the cooperation between the five activities of the supply chain. To illustrate how this would be achieved, consider an example.

Suppose a large corporation located in Paris is in the process of upgrading its computer hardware throughout the organization. For this purpose, they need 3000 PCs of a particular model. After technical evaluation, they have shortlisted three models from three different manufacturers. The corporation first approaches the retailer for the manufacturer that is its first preference and that has an office in Paris. The client understands that such a large order cannot be bought off the shelf and, hence, is willing to wait for up to 30 days to receive the entire order. Although the retailer has a total inventory of 5000 PCs, it has only 250 pieces of that particular model in store and based on historical demand is expecting to receive 250 more in 15 days. The retailer contacts its partner retailers in Brussels, Frankfurt, and Zurich, but these can provide only an additional 600. The manufacturer has manufacturing facilities in the United States and Asia. The retailer in Paris traditionally receives its supplies from the U.S. manufacturing plant by sea. Normally, it takes 90 days between the time the order is placed and received. However, an express order can be supplied in 60 days. These 60 days are broken down as follows:

4.3. BUILDING AN EFFICIENT SUPPLY CHAIN AT THE TACTICAL LEVEL 71

It takes 2 or 3 days for the retailer to process a special order of such a large quantity because it requires attention of several layers of management. It takes 2 or 3 days for the order processing department in the United States to enter the order in their system, plan for it, and then release the order to the manufacturing facility. The manufacturing facility runs its MRP planning system every week; hence, up to a week may be needed for the order to be planned. Once the MRP run is complete, it sends the component requirement to the procurement division, which then takes a couple of days to process it through its system and inform the suppliers. Fortunately, the manufacturer has a very good understanding with the suppliers and can supply the components within 7 days. Of these, a couple of days are spent processing the order and planning it.

Traditionally, it takes 7 days to complete an order once all raw materials become available. However, shop floor management is allowed to take 14 days to complete an order to allow for the day to day disruptions due to machine breakdowns, worker absenteeism, inventory reconciliation in different stages, quality problems, etc.

It takes 2 weeks to ship the order to France. It takes another week for the consignment to be delivered from the port to the store in Paris, receiving and reporting it in inventory and becoming available for customers.

In addition, there is excessive demand for PCs in the U.S. market because many companies that were waiting for the Year 2000 date problem to resolve are now buying PCs in large quantities. Thus, the U.S. manufacturing facility is already running a backlog in fulfilling the demand. Given this scenario, this supply chain is about to lose a large order despite the close coordination between the retailer, manufacturer, and the component suppliers.

Now consider that the supply chain has an information system similar to that shown in Fig. 4.3. The retailer now has a view of the current status of all other partners. It sees that due to low demand in Asia, the manufacturing plant there has much slack capacity. The manufacturing plant and the suppliers in Asia have a close relationship similar to that in the United States. Eastern Europe is supplied by the plant in Asia and there is a shipment leaving for Athens, Greece, in 14 days. Thus, the retailer places the order with the Asia plant. The moment the order is placed, it is seen by the suppliers and the manufacturing plant (the time spent in processing the order in multiple steps in the supply chain is thus eliminated), which plan for them accordingly. The raw materials reach the manufacturer in 5 days and the order is completed in 7 days, just in time for the shipment to Athens. The order is then transported on trucks directly from the port to the customer well within the 30-day required time.

This example highlights the strength of close cooperation between the activities of buy, make, move, and sell in a supply chain (almost eliminating the need for the store activity) as enabled by an information system shared by these partners. The availability of the system enabled meeting customer demand in a quick time frame without the need for carrying expensive inventory by any partner of the supply chain or incurring extra costs in expedited shipments (such as air freight). This example highlights how a customer order can be satisfied in the shortest possible amount of time at minimum cost. This requires determining whether

- The order can be satisfied by existing inventory at a retail location near the customer; if not, can it be satisfied by inventories at other locations without incurring excess transportation costs?
- The order can be satisfied through manufacturing; if yes, which manufacturing plant should make it?
- The partner responsible for the move activity can add value to the supply chain by combining different orders in one shipment, thereby reducing cost and decreasing the transportation time. It must also be determined whether it can handle special requests such as deliveries directly to customers (and not just between the manufacturer, wholesaler, and retailer).

Furthermore, it may be possible to reduce the time needed to perform each of the activities corresponding to the project under consideration. This can be done either gradually or abruptly. Gradual reduction of time is achievable through the mapping of the project process. This mapping leads to a graph similar to the one shown in Fig. 4.5, in which "cost" is replaced by "time." The reduction of time is then based on activity analyses that should preferably be performed by the people in charge of this activity. The analyses often lead to suppression of non-value-adding times, such as reduced storage time, reduced setup times as demonstrated by the single minute exchange of die method, or reduced processing time by modifying the processes used to perform some activities.

Reducing the processing time of a project can sometimes be done through reengineering, which consists of changing the whole process and even some of the resources. This abrupt approach is not well accepted by the project team and may lead only to a partial success or, sometimes, to a disaster. Similar approaches can be used for quality improvement and cost reductions and are discussed later.

4.3. BUILDING AN EFFICIENT SUPPLY CHAIN AT THE TACTICAL LEVEL

> Minimizing the time required to convert orders into cash requires scheduling each order as soon as it arrives in the supply chain. This scheduling activity should simultaneously cover all activities (i.e., buy, make, move, store, and sell), taking into account the partial schedule of the orders that previously arrived in the supply chain. Due to the complexity of scheduling problems and the fact that each order is scheduled as soon as it arrives in the supply chain, rescheduling of existing tasks should be avoided, except in exceptional cases.
>
> The time needed to perform the activities required to complete a process can be reduced gradually through the mapping of the project process and analysis of each of the activities. Reengineering is an abrupt approach that is difficult to apply.

4.3.2.2. Minimizing the Total Work-in-Process

In the past, the relationships between the partners of the production chains were more adversarial than cooperative. Each one was trying to increase its own efficiency instead of working to increase the global efficiency of the system. As a consequence, the goal of each partner was not to decrease inventories, since inventories favor productivity, but to move them upstream or downstream in the chain in order to keep the advantages of inventories while transferring the related costs to other partners or subcontractors. This was quite common in the auto industry, in which auto makers used to ask subcontractors to deliver parts or subsystems "just in time," which often resulted in transferring inventories to subcontractors. It is still the case in production systems working on a master–slave basis: The "master" builds its success at the expense of the "slaves."

The philosophy behind the supply chain paradigm is totally different: The goal is to improve the efficiency of the whole system and thus to reduce the total work-in-process (WIP). Figure 4.4 illustrates the difference between the past and the current objective for WIP.

The approach suggested in the previous section that consists of scheduling an order as soon as it arrives in the supply chain can be improved by considering storage facilities as resources having flexible but upper bounded manufacturing times. These manufacturing times represent the maximal period a product can be stored in the corresponding facility. Doing so, we can reduce the WIP by reducing the upper bounds of the manufacturing

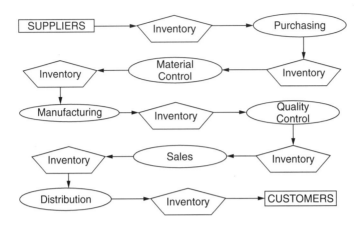

a. Former Situation

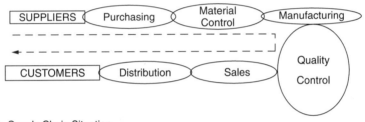

b. Supply Chain Situation

FIGURE 4.4 WIP in a supply chain.

times associated with the storage facilities at the expense of the reduction of productivity. Minimal productivity is reached when the upper bounds of the manufacturing times associated with the storage facilities are equal to zero. In this case, the WIP is reduced to the products on which operations are performed: Such a schedule is referred to as a no-wait schedule.

> Global real-time scheduling can minimize WIP. One must consider each inventory as a resource having a limited "manufacturing time" to force each product to limitits stay in the storage facility.

4.3.2.3. Improving Pipeline Visibility

The communication network is the nervous system of the supply chain. The evolution of computer systems from central mainframe to local workstations drastically changed the way information is delivered. Nowadays, information can be sent from the place it is generated to any other place in

4.3. BUILDING AN EFFICIENT SUPPLY CHAIN AT THE TACTICAL LEVEL

the supply chain in real time. Theoretically, this allows close monitoring of product movements, inventories, market changes, logistics, etc.

In other words, the current state of the supply chain, as well as the ongoing moves of products and information, can be obtained at any time. It is important to note that bar codes have made it possible to precisely track product movements in real time and with a low rate of error.

Thus, technical barriers in information systems have been virtually removed. However, two problems remain when a new supply chain is implemented: The selection of information to be sent to each partner, including when the information should be sent, and the removal of the psychological barriers that incite people and organizations to hold back information.

A systematic analysis can be conducted to define the information needed by each of the partners as well as the circumstances in which this information must be provided. It is also possible to define the best form in which to present this information (ergonomics) and the partner that will be in charge of generating it.

Removing the psychological barriers is more complicated, especially in a situation in which sharing of information is considered a way of losing power or endangering competitiveness. The most difficult challenge encountered when organizing a supply chain is to convince the partners that the success of each of them lies in the success of the supply chain as a whole.

At the highest management level or the strategic level, the acceptance of an open cooperation is usually quite easy, assuming that the whole mechanism of the system is clearly defined and that the roles and responsibilities of each partner are precisely specified and mentioned in contracts.

It is much more difficult to convince people who are acting at the tactical or operational level, especially if they have different backgrounds. Nevertheless, various actions can be taken at the tactical level to achieve this goal, including (i) training employees of the different partners together; (ii) assigning leading employees, for a short period, in the departments (or partners' companies) with which they are associated in their daily activities; and (iii) introducing supplementary benefits that are linked to the success of interactivity cooperation. The goal is to help employees converge toward a common way of conducting the global management of the supply chain and extending their own management horizon.

If designed and managed appropriately, the information system should provide each employee with any information he or she needs to perfectly perform his or her tasks in any circumstance. Pipeline visibility (i.e., the ability to obtain any information wherever and whenever needed) is a necessary condition for an efficient supply chain, but it must be supplemented with an effective management system.

An information system is the tool for tactical and operational management, who work with the set of rules that are supposed to optimize the behavior of the whole system, taking into account the constraints and the objectives prescribed by high-level management. Pipeline visibility is the necessary condition for implementing successful real-time management, as mentioned in Section 4.3.2.

> Due to the efficiency of current information systems (see Chapter 6), obtaining pipeline visibility is theoretically an easy task. Nevertheless, it requires solving two difficult problems: (i) selecting and managing information to be sent to each partner in real time and (ii) supporting the required psychological and behavioral changes of the employees.

4.3.2.4. Improving Visibility of Demand

An efficient supply chain is supposed to be responsive and fast to react quickly to customers' requirements. In the past, forecasting was the method to manage production at the strategic and tactical levels. Forecasting consists of extrapolating the past behavior of customers. In this kind of approach, it is assumed that customers' requirements in the future can be derived from the corresponding values in the past taking into account their tendencies. It is well-known that this hypothesis does not hold anymore. In other words, it is no longer possible to "manage with the driving mirror." This is due to technological innovations and the ever-changing requirements of the customers. As a consequence, we have to introduce methods that capture the wishes of the customers as soon as possible and even before customers are aware of their own wishes.

One of the popular methods that can be applied is the so-called Vendor Managed Inventory. In this method, retailers are supposed to inform the partners of the supply chain, in real time, of the sales, evolution of the inventories, marketing activities such as promotion and participation in shows, and any customer behavior that may influence future buying habits. This method is a major step forward. Nevertheless, the underlying assumption in this method is similar to that of the forecasting method: the continuity of the buying habits of the customers. The only improvements are the real-time aspect of the method and the information on marketing activities, which may help to forecast demands.

A further step forward in improving demand visibility consists of detecting customers' requirement as soon as they emerge in their brains and even before. To reach this goal, it is necessary to know precisely the demographic and psychological profiles of the customers and thus be in close contact with them. The profiles are derived from their past buying history, the information concerning their buying capacity (i.e., their income level), and social position in cases in which customers are individuals. A classification can be done based on their profiles. As a result, we obtain customers' classes, with each class being defined by the type of demands that may be made by the elements of the class. A good understanding of the technological trends combined with the knowledge provided by these classes can lead to the functional specification of new products and their importance on the market. It should be noted that valuable information can be obtained from some Internet providers who trace the connections of their subscriber and thus can help to design their profiles.

> To conclude, the goal when improving the visibility of demand is to move the demand penetration point — that is, the point of the supply chain at which the demand is known — as downstream as possible. Tools are available to determine with a reasonable probability the customers' demands even before they emerge in their brains.

4.3.2.5. Improving Quality

Quality is often defined as the set of properties and characteristics of a product or a service that allows it to meet requirements that are explicitly or implicitly expressed by the customers. This definition stresses the fact that a need could be identified by the supply chain before it is clearly expressed by the customers. In other words, this definition takes into account the fact that the demand penetration point may have been pushed very close to the beginning of the pipeline, as explained previously.

Three main aspects should be considered with regard to quality: quality mastery, quality assurance, and total quality.

Quality Mastery

Quality mastery involves evaluating the product or service characteristics' fit with the specifications provided by the designers or the customers. Quality mastery implies the ability to measure quality, which in turn allows measuring the efficiency of the activities performed to improve quality. A tool widely used to complete quality mastery is a standardized table that distinguishes between internal and external dysfunction costs as well as between the costs to prevent a dysfunction and the costs to correct it. Such a standardized table, called cost for quality (CQ), helps the user identify the actions to be taken to reach the required quality.

CQ tables depend on the type of company. The internal dysfunction costs usually include the costs for scrap, sorting, reworking, retesting, rescheduling, and redesigning. External dysfunction costs usually include costs related to warranty, complaints, ineffective deliveries, allowance, and incomplete documentation.

Internal and external actions to be taken to prevent and correct internal and external dysfunction are then specified and their costs are evaluated. For instance, scrap problems can be partially solved by taking the following internal preventive actions:

- Redesigning products (partially or totally).
- Modifying manufacturing processes. This may be done while introducing more efficient resources.
- Training employees.
- Improving quality control by introducing new tests and/or by changing the testing points.
- Introducing automation at some production levels.

Internal actions to correct scrap may include design activities to reuse scrap for other products or the same type of products.

External actions to prevent scrap include

- Checking and evaluating providers of raw material and components to select the best ones.
- Reducing the number of providers, which may lead to a risk of running out of raw material or components: A cost is associated with this risk.
- Training suppliers.
- Analyzing customer requirements to determine if it is possible to simplify the design and the production. In other words, try to introduce a new generation of products that are easy to manufacture and thus lead to a drastic reduction of scrap.
- Outsourcing specific operations that require special capabilities.

Finally, external activities to correct scrap may include the search of companies that may valorize scrap.

Note that the actions preventing or correcting scrap may also prevent or correct other dysfunctions. For instance, switching to another generation of products may also reduce complaints and the costs related to warranty.

4.3. BUILDING AN EFFICIENT SUPPLY CHAIN AT THE TACTICAL LEVEL

TABLE 4.1 Solving the Scrap Problem

	Prevent	Correct
Internal activities	Redesigning products Modifying management process Training employees Improving quality control Introducing automation	Reusing scrap
External activities	Selecting providers Reducing the number of providers Switching to next generation of products Outsourcing	External valorization of scrap

TABLE 4.2 Reducing the Number of Complaints

	Prevent	Correct
Internal activities	Improving documentation Training customers	24-hr-a-day support service Hot line
External activities	Training retailers Advertising	Dissemination of advisers throughout the country

Actions to take in order to solve the scrap problem are shown in Table 4.1. The same type of table is introduced for each one of the external or internal dysfunctions. For instance, complaints may lead to the development of Table 4.2. Each one of the actions listed in the table is then evaluated in terms of the cost to conduct the action and the benefit resulting from the removal of the dysfunction. As mentioned previously, the same action may prevent or correct several types of dysfunction. It is thus reasonable to group all the cost and benefit information in the same table. Table 4.3 is an example of a cost table.

To summarize, one table is attached to each internal or external dysfunction (Tables 4.1 and 4.2). The goal of these tables is to establish the checklists of the actions to be taken to solve the dysfunction problems. The costs and benefits are then evaluated and gathered in a table similar to Table 4.3.

In Table 4.3, A_1 to A_r are the preventive actions that are taken. They are followed by evaluation of their cost. Similarly, B_1-B_z are the corrective actions that are followed by the evaluation of their cost. ID is the set of internal dysfunction D_1-D_k to be prevented or corrected, whereas ED is the set of external dysfunction E_1 to E_s. The elements of each row of the

TABLE 4.3 Cost Table

		PREVENTIVE ACTIONS				CORRECTIVE ACTIONS				
		A1 + cost	A2 + cost			Ar + cost	B1 + cost	B2 + cost			Bz + cost
I. D.	D1										
	D2										
	Dk										
E. D.	E1										
	E2										
	Es										

table contain the reduction of expenses expected for the dysfunction when the different actions are taken. Note that these amounts will depend on the order in which actions will be taken.

Quality Assurance

The goal of quality assurance is to guarantee the required quality level for products and services. Quality assurance is often supported by the International Standard for Organization (ISO). ISO 9004 provides a guide for the management of quality system, whereas ISO 9001, 9002, and 9003 aim to establish quality standards that guarantee the level and invariability of quality.

Total Quality

Total quality requires a special management, called total quality management (TQM), that involves each employee in the organization in the improvement of quality. TQM is customer and project oriented. Being customer oriented means that the efforts of the employees are directed toward customer satisfaction, which is also the goal of supply chains as we define them. Being project oriented refers to the fact that the system at hand is viewed as a set of processes that are evaluated globally but independently from each other. This is also a characteristic of supply chains. TQM can be applied using different methods. The most popular ones are the Kaizen method and reengineering.

The Kaizen method provides a gradual and constant improvement of quality. It can also be defined as a continuous effort made by each employee to use the human and physical resources for quality improvement. In the Kaizen method, employees are encouraged to make proposals to improve the processes at their own level of responsibility.

Reengineering consists of reconsidering the ways projects are performed in order to improve quality and reduce costs. Reengineering is not gradual, as is the Kaizen method, but fast and abrupt. It reconsiders the processes from scratch, introducing possibly more efficient resources. Reengineering is often the consequence of tough competition that forces a company to develop products or services that offer better functionality in a short amount of time and at the lowest cost.

Some rules have been proposed by the car company Renault for successful reengineering. The most meaningful are the following:

- Think in terms of projects and not of functions.
- Involve people who will use the outputs of the project (product or service) in the design of the process.
- Ask employees who will be in charge of performing the tasks to determine the design of the information system.
- Locate the decision-making system where the work is performed.
- Integrate controls in the process.
- Take each piece of information only once, at the place it is generated.
- When several operations can be performed in parallel, define in advance the rules to apply to schedule these operations.

As mentioned earlier, due to the abruptness of reengineering, employees often resist it, which results in a working system that is not exactly what was initially expected and can have dangerous consequences. Therefore, reengineering should be used only if necessary.

Quality mastery, quality assurance, and total quality should be tailored for supply chains, assuming that suppliers and even customers share a common quality approach — that is, agree on the way in which quality mastery is conducted, use the same ISO environment for quality assurance, introduce the same Kaizen philosophy, and closely cooperate if reengineering is necessary. Furthermore, when a lack of quality arises, the problem should be analyzed, if possible, in collaboration with the suppliers and even with customers.

> Quality is a necessary condition for a successful supply chain. It is a goal that should be reached in close collaboration with suppliers and even customers. Quality mastery aims at evaluating the quality. Quality assurance guarantees a constant and well-defined quality level to the partners. Total quality is a never-ending effort to improve quality. It could be gradual (Kaizen) or drastic (reengineering).

4.3.2.6. Reducing Costs

Being able to measure profitability is a key issue in any production system. For years, so-called analytical accounting relied on arbitrary allocation of indirect costs to product types or services, and hence made it impossible to evaluate the true profitability of these product types or services. Accounting techniques have evolved dramatically (and positively) during the past few years due to the new project-oriented paradigm, which is the most important characteristic of supply chains.

Supply chain management is flow oriented, which means that a supply chain is no longer viewed as a set of departments specialized in some activities but, rather, as a set of projects, with each project having a specific objective suchas manufacturing a given type of product or providing a well-defined service. To reach this goal, each project uses the facilities provided by the supply chain, ensuring that no barriers between the activities delay the completion of the project. Cost analysis should include each of the projects. More precisely, there should be a one-to-one relationship between the projects and data cost. Indeed, cost evaluation could be detailed to different levels. A cost can be evaluated for each activity performed in the project, but usually activities are grouped into activity centers, and a cost is evaluated for each activity center that takes part in the project under consideration. In some cases, projects are categorized in accordance with customer segments, which provides a more detailed view of the costs related to the projects or types of product.

Another important aspect of cost evaluation in supply chains is the concept of incremental costs. According to this point of view, only the additional costs incurred by the activity centers to perform a project are taken into account in the evaluation of the project. Indeed, the resulting cost will

4.3. BUILDING AN EFFICIENT SUPPLY CHAIN AT THE TACTICAL LEVEL

depend on the projects already performed in the supply chain. For instance, when launching a new type of product, a supply chain may take advantage of some underloaded resources already available. This would lead to a low incremental cost. On the contrary, the same type of product launched in a supply chain in which none of the existing resources can be used would lead to a high incremental cost. Usually, this problem can be neglected due to the reasonable spectrum of resources that are typically present in a supply chain.

If we want to evaluate the incremental costs resulting from the manufacturing of a given type of product for a given customer segment, we have to answer the following question: What is the cost per unit that could be avoided in the following activities if we stop this production?

- Design and redesign
- Order processing
- Raw materials and components
- Marketing
- Communication
- Documentation
- Transportation
- Manufacturing, which includes salaries, energy consumption, maintenance, wear of machines, quality control, and material handling
- Rework, in the case of lack of quality
- Management
- Specific storage facilities
- Promotional activities
- Packaging
- Delivery

Note that these costs should be recorded when the corresponding activities are performed.

Table 4.4, represents the way in which cost evaluation should be done in a supply chain. $P_1, P_2, \ldots, P_k$ are the projects or products being worked on in the supply chain; $P_{i1}, P_{i2}, \ldots, P_{ki}$ are the parts of P_i that are decomposed in accordance with customer segments; and $AC_1, AC_2, \ldots, AC_n$ are the n activity centers that comprise the supply chain. The costs associated with the projects and the activity centers that are not represented in the table are supposed to be equal to zero. This kind of cost analysis may lead to a decision to stop a given production activity or service for a specific type

TABLE 4.4 Project Evaluation

		AC_1	AC_2	...	AC_n	Total
P_1	P_{11}	20	15		30	65
	P_{12}	10	25		15	50
P_2	P_{21}	20	30		10	60
	P_{22}	25	5		18	48
	P_{23}	10	10		5	25
⋮						
P_k	P_{k1}	17	3		9	29
	P_{k2}	4	7		12	23
Total		106	95		99	300

of customer or, on the contrary, to introduce a new production activity or service in the supply chain.

Indeed, in the incremental evaluation described previously, we assume that each project (product type or service) is performed optimally (i.e., at the lowest cost). This implies that the following questions have been answered:

- For each activity required to complete the project, is it possible to reduce the cost by using other resources or changing the way the activity is performed (Kaizen-like approach)? If the answer is "yes," what would be the cost?
- Is it possible to perform the same project using another process that is cheaper than the current one (reengineering)? If the answer is "yes," what would be the cost?

The tool often used to analyze and possibly reduce the costs of a project is a graph in which each of the activities is represented with its cost, evaluated as explained previously. Figure 4.5 represents such a graph in which the project consists of manufacturing a type of product. To reduce the size of the graph, we represent five activity centers instead of the detailed activities. Buy, make, move, store, and sell denote these activity centers.

The following basic rules are taken into account when evaluating the costs in a supply chain:

- Costs should be attached to projects instead of departments (i.e., the approach when evaluating costs should be horizontal instead of vertical).

4.3. Building an Efficient Supply Chain at the Tactical Level

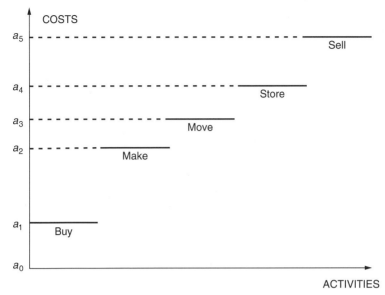

FIGURE 4.5 The activities–incremental costs graph.

- Only incremental costs should be considered. These incremental costs should be evaluated for each activity of the project and even for each customer segment.

  The following approaches are used to reduce the incremental costs of a project:

- A gradual approach, similar to the Kaizen approach applied to improve quality
- A more abrupt approach that consists of changing drastically the manufacturing process (reengineering)

4.3.2.7. Improving Services

Another important objective is to maximize the evaluation of supply chain performances as made by customers. To achieve this goal, managers should carefully check the following aspects:

- *Time from customer orders to deliveries:* This time must be reduced as much as possible. This aspect is closely connected with the efforts made to minimize the time required to convert orders into cash (see Section 4.3.2.1).

- *Quality of documentation:* A customer must be able to find easily and quickly, in the documentation, any information needed to obtain the best possible service from his or her acquisition.
- *Quality of customers' reception:* The reception office is the window of the supply chain. The first contact of the customer with the system is made through the reception office. Therefore, this activity should be placed in the hands of carefully selected employees. In the world of Internet-based retailing, this translates into having an appealing web site that is attractive but at the same time easy to use by the customer.
- *Information to customers:* The supply chain should be able to provide any relevant information, at any time, to the customers. This includes information related to the state of their orders, the state of products under repair, the state of services that are in preparation, or simply information about products or services in which a customer may be interested.
- *Reaction time to complaints:* This is a very important factor for customers. It goes hand in hand with the quality of the technical support. The goal is to minimize the time the service is interrupted either by repairing or replacing the unusable product or by providing an identical product for use during the time in which the repair is performed. If the supply chain provides a service, the goal is to correct the dysfunction as fast as possible.
- *Flexibility of the supply chain:* The more flexible the supply chain, the easier it is able to adapt to variations in customer demands. Flexibility can be measured by considering the width of the spectrum of the system outputs. In this new environment of globalization, customer needs are becoming increasingly specific. Customers from different regions of the world have different likings and tastes. Even simple products such as blenders may have to be customized for local needs (such as government regulations, technical requirements, and aesthetics appealing to different demographics). The supply chain has to be flexible enough to cope with the demand from different markets it serves and quickly scale up or ramp down production for different markets at low costs.
- *Maintenance quality and efficiency:* Maintenance should not only be well done but also disturb the customers as little as possible. Maintenance is an activity that must be taken into account at the design level in order to reduce as much as possible the maintenance periods. An example is maintenance of cars that

requires less time and that is currently necessary every 15,000 km instead of every 10,000 km, 3 or 4 years ago.

The production or servicing activities can be improved by comparing them to similar activities not only in competitors' supply chains but also in supply chains that act in totally different domains. For instance, a supply chain devoted to car manufacturing may improve its transportation system for better customer satisfaction by analyzing how the steel industry or the pharmaceutical industry organizes its transportation system.

> Factors that improve customer service are reduction of time from customer orders to deliveries and improvement of the quality of product/service documentation, reception of customers, and information about the supply chain. Another important factor is a fast and effective reaction to complaints. Flexibility, which is the ability to adapt the outputs of the supply chain to the evolution of customer requirements, is a major factor for competitiveness.

4.4. PERFORMANCE EVALUATION OF A SUPPLY CHAIN

Performance measures to evaluate a supply chain should cover the financial and operational domains since the goal is to provide customer satisfaction at low cost and to guarantee competitiveness over the long term. In other words, performance measures should be useful not only to continuously improve the efficiency of the supply chain but also to help run a strategic policy. Performance measures should be easy to define, simple to apply, and easy to understand in order for the manager to be able to react accordingly in real time in day-to-day situations.

Financial performance reflects the profitability of the system and its ability to be competitive in the long term. Operational performance is one of the factors that lead to customer satisfaction. Evaluation of these two aspects is described in detail next.

4.4.1. FINANCIAL EVALUATION

Any decisions made in an activity influence all other activities. As a consequence, as mentioned in Section 3.3.2.6, cost evaluation should be connected to projects instead of departments. In other words, costs should be closely linked to the flows of activities that lead to the final products or services. Furthermore, only incremental costs should be considered in order to avoid arbitrary assignment of fixed costs to projects. We also mentioned that the flows of activities should be attached to activity centers.

Costs to take into account in the evaluation are the costs per unit (of product or service) and costs per activity center that could be avoided if production of the type of product or service under consideration was halted.

For instance, assume that we are interested in the costs incurred when producing a given type of refrigerator. The following question must be answered: What would be the cost per unit that could be avoided if we stop manufacturing this type of refrigerator in the following activity centers?

- Buying, in which buying, handling, and storing of raw material and components dedicated to this type of refrigerator occur
- Manufacturing, which could be further divided into more detailed activity centers such as assembling, testing, and painting
- Packing, in which preparing the dispatching of the refrigerators occurs
- Delivering to retailers
- Controlling quality
- Managing the project

The cost per unit avoided should be compared with the price paid by customers to enjoy the product or the service.

Financial evaluation also encapsulates the measure of non-value-adding expenses. The following parameters are important from this standpoint:

- How long does it take, on average, to send an invoice to a customer after sending the order?
- What is the percentage of inaccurate invoices?
- What is the percentage of incomplete deliveries?
- What is the percentage of returned delivery? A delivery is returned when it is late, when it contains defective products, or when it does not fit with the order.

These percentages should preferably be measured in monetary terms.

The previous evaluation tasks concern the short-term horizon. This is not sufficient to decide if a product type or service should be a part of the supply chain activity. We have to integrate three more parameters in our evaluation, namely, R&D activities, financial resources utilization, and cash flow.

R&D activity, which is usually very expensive, is necessary to preserve the competitiveness of the supply chain in the long term. It is usually difficult, if not impossible, to assign a part of the R&D cost to a specific type of product or service, except if the R&D activity is dedicated to the project under consideration. This problem is a drawback of supply chains. In many companies, changing from a vertical (department-oriented) to a horizontal (project-oriented) organization has given rise to difficulties and even to a drastic reduction of the R&D activity, and thus to a weakness of the company in the medium and long term. In other words, the introduction of supply chains has often led to a focus on the short term, which handicapped the future competitiveness of the systems. In fact, it is very difficult to integrate R&D activities in the projects in most cases. Two approaches can be used to address this problem, but neither is fully satisfactory. The first one consists of continuing the traditional R&D organization and considering it as an "internal subcontractor" for the projects developed in the supply chain. In this case, conflicts may arise between the project leaders, mainly because the timescale is very short for projects compared to that for R&D. Another problem is the definition of the cost of an R&D service since most of the studies and tools (software and hardware) developed may be common to different projects in a supply chain. The second solution is to subcontract R&D activities. The drawbacks in this case are cost, which is usually very high, and confidentiality. In practice, a combination of these two approaches works better: Sensitive R&D activities are kept in-house while minor developments or fundamental researches are performed outside the supply chain.

Financial resource utilization is another important aspect when evaluating a project since one of the goals of evaluation is to maximize the productivity of capital in order to attract investments. Thus, return on investment (ROI), which is the ratio of profit earned over capital required to perform the project, is an important criterion to take into account when evaluating a project. ROI is sometimes expressed as follows:

$$\text{ROI} = \underbrace{(\text{profit/sales})}_{\text{margin}} \underbrace{(\text{sales/required capital})}_{\text{capital turnover}}$$

Written as the product of margin by capital turnover, ROI is not only more understandable but also can be easily optimized because profit and sales, on the one hand, and sales and required capital, on the other hand, are handled at the same level of responsibility in companies. Thus, ROI can be optimized much more easily.

Another important parameter to consider is cash flow. Simply stated, cash flow is the capital available for investments. It represents the ability of a supply chain to seize new opportunities. The stronger the cash flow, the more reactive the supply chain. Having a strong positive cash flow is thus another important goal when managing a supply chain. As a consequence, the global evaluation of any project that would free a large amount of cash flow if discontinued will be highly scrutinized.

> In the short term, financial evaluation consists of measuring the incremental cost per unit in each activity center and for each project, and measuring the non-value-adding expenses.
>
> In the medium and long term, R&D is difficult to integrate in a supply chain because R&D cannot be split in order to fit with each of the projects individually and the horizon of a project is usually much shorter than that of R&D activities. Managers should also keep in mind that investors want to maximize the productivity of capital, which advocates for maximizing margin and capital turnover. They must also preserve the flexibility of their strategic decisions by preserving a significant cash flow.

4.4.2. OPERATIONAL EVALUATION

Customer satisfaction is the ultimate goal of a supply chain; hence, it is a measure of operational excellence in the supply chain. Various parameters should be measured to evaluate this aspect.

4.4.2.1. Availability

This parameter measures the ability of the supply chain to make products or services available to customers faster than can competitors. This order-fulfillment rate or time-delivery parameter is one of the more established performance measurement criteria. In the case of off-the-shelf products, availability should be computed as the number of times the product was available on the shelf when a customer wanted it. In traditional supply chains, a 100% availability ratio or customer fill rate is difficult to achieve for a wide variety of products (especially those with many variations or options) without driving the inventory levels to infinity in the case of off-the-shelf products. To ensure high availability, good demand forecasting is necessary. More important is the ability to replenish out-of-stock products. A related issue is how soon the retailer realizes that it is out of stock. These issues are best addressed with an integrated information system as discussed previously.

4.4. Performance Evaluation of a Supply Chain

For make-to-order products, the customer expects to wait a certain amount of time after placing the order to receive the delivery. In such cases, the competitiveness of the supply chain is determined by the time it takes to fulfill the order. An associated measure is on-time delivery, which computes the percentage of time the correct product was delivered on the promised date.

4.4.2.2. Adequacy of Customer Expectations

The question to be answered here is the following: Are the products or services being provided closer to customer requirements than those of the competitors? The comparison of the "closeness" of products or services to customer expectation is difficult to measure. Closeness to expectation includes not only technical aspects but also the perception that the customers have of the ease of use of a product, the efficiency of the service provided, the aesthetics of the product, and the quality of the product. Customer perception is of utmost importance and often determines the success or failure of a product or a service. Thus, this aspect should be handled carefully.

Consider the case of a supply chain providing products. Indeed, the first characteristics to evaluate are the technical ones. The technical specifications of the product must correspond to those included in the documents that advertise it. However, this is not sufficient since, even for high-tech products (e.g., computers), the perception of the product changes from one person to another. This perception depends on the background of customers, the kind of application they anticipate, their understanding of the documents related to the product, the ease with which they can use the product, etc.

To determine the subjective perception that potential customers have of a product, an efficient approach is to create a poll and to gather for each customer interviewed selected characteristics of the customer and his or her perception of the product. Selected characteristics of customers include their general background, their knowledge about similar products and the domain in which the product is used, and, more generally, any information that may predict the reaction of the customers when using the product. The perception of the product includes the judgment of the customer regarding the adequacy of the product with what he or she was expecting, the judgment of the quality of the product, the level of service received, the ease of use of the product, and, more generally, any judgment of the customer on the product.

Next, a rectangular table, called a contingency table, is constructed with as many rows as the number of selected characteristics of the customers and as many columns as the parameters evaluated. Table 4.5 is an example of a

TABLE 4.5 Example Contingency Table

Customer Perception	Quality			Robustness			Efficiency			Easy to Use		Easy to Maintain			Cheap			First on the Market	
	High	Medium	Low	High	Medium	Low	High	Medium	Low	Yes	No	Yes	Medium	No	Yes	Medium	No	Yes	No
Customer characteristics																			
Professional																				
Nonprofessional																				
High income																				
Low income																				
Age																				
20–30																				
31–40																				
41–50																				
Over 50																				
Income																				
$30,000																				
$30,000–40,000																				
$41,000–60,000																				
>$60,000																				
Male																				
Female																				

contingency table. At the intersection of row i and column j, we indicate the number of interviewed customers having characteristic i who expressed a judgment j about the product. Correspondence analysis, a technique from data analysis (a branch of statistics), is then applied to this rectangular table. The result of this process is to connect the characteristics of the customers with their subjective perception of the product. The result is a set of two-dimensional graphs that show customer characteristics that are close to customer perception. For instance, we may find that a perception of low quality is related to professionals with annual incomes higher than $60,000. If this segment is of interest for the company, efforts will be made to improve the perception of quality. As a result, we are able to perform segmentation of the market by defining the type of customer that has a positive evaluation of the product. This knowledge can then be used to align product features more closely with different market segments.

It is also possible to conduct the same analysis by replacing the characteristics of the product under consideration by the characteristics that a similar future product could have. This would allow defining the type of customers that would be attracted by this product. Continuing these correspondence analyses would provide the results to complete the segmentation of the market.

Correspondence analysis can also be used to determine the perception that customers have of a service. For instance, characteristics of services include the adaptability of services to customer requirements, their quality, and their spectrum.

4.4.2.3. Customer Service

Measuring and evaluating customer service is of utmost importance. The following measures should be made:

- How many phone calls are necessary, on average, to reach an employee? The result should be as close as possible to one.
- How many employees should the customer contact in order to obtain the information he or she needs? The result should also be as close as possible to one.
- How long does it take to obtain an appointment?
- How long does it take to replace the product in case of breakdown or to replace the provider of the service if necessary?
- What is the average delivery delay? This aspect is important to gain new customers and keep them. As a consequence, this delay should be equal to zero.

- What is the percentage of first-time satisfied orders (i.e., orders that are correct in quantity and quality)?
- What is the percentage of invoices that are not accurate?
- What is the time required, on average, to complete an incomplete delivery?
- What is the time required, on average, to correct an invoice that is not accurate?
- Does the supply chain offer training if necessary?

Results from these measures should be tabulated to collect these metrics and scores. These are then used to determine the overall performance of the supply chain. Weighted sums should be used to represent the relative importance of each of these parameters.

> Customer satisfaction is the goal to reach in order to attract and keep customers. The evaluation of the supply chain in achieving this goal considers the availability of the products and services and their adequacy to customer expectations. Another important aspect that should be evaluated is customer service. The following aspects should be measured: ease of contacting the partner in charge of customers, the quality of the delivery of orders, the reactivity of the system to customer complaints, etc.

4.5. CONCLUSION

Supply chains are production systems that are project oriented; in other words, they are systems that consider production from an horizontal standpoint instead of a vertical one. In doing so, a supply chain breaks the departmental walls and barriers and thus allows faster flow of products and services, making the entire system much more efficient. Supply chains also make production easily controllable and simplify performance evaluation. It should be noted that only incremental costs are taken into account in order to avoid allocating indirect costs arbitrary to projects. In this chapter, we addressed the tactical level of the supply chain and stressed the importance of real-time scheduling.

In making decisions at the tactical level, managers should consider the following: First, tactical decisions should be in line with the strategic level. They should not overlook the strategic goals (e.g., short-term profits should not sacrifice long-term benefits to be derived from strategic activities such as R&D). Second, managers should avoid conflicts with other partners in the supply chain within the framework of the strategic arrangement.

REFERENCES

Aitken, J., "Supply chain integration within the context of a supplier association," PhD thesis, Cranfield University, Cranfield, Bedfordshire, UK, 1998.

Anand, K. S., and Mendelson, H., "Information and organization for horizontal multimarket coordination," *Management Sci.* **43**(12), 1609–1627, 1997.

Arntzen, B. C., Brown, G. G., Harrison, T. P., and Trafton, L. L., "Global supply chain management at Digital Equipment Corporation," *Interface* **25**(1), 69–93, 1995.

Ballon, R. H., *Business Logistics Management*. Prentice Hall, Englewood Cliffs, NJ, 1992.

Bleakley, R. R., "Strange bedfellows: Some companies let suppliers work on site and even place orders," *Wall Street Journal*, A1, January 13, 1995.

Braithwaite, A., and Christopher, M., "Managing the global pipeline," *Int. J. Logistics Management* **2**(2), 55–62, 1991.

Cavinato, J. L., "Identifying interfirm total cost advantages for supply chain competitiveness," *Int. J. Purchasing Materials Management* **27**, 10–15, 1991.

Christopher, M., *Logistic and Supply Chain Management. Strategies for Reducing Cost and Improving Service*, 2nd ed., Financial Time Management, London, 1998. [ISBN 0-273-63049-0]

Cooper, R., and Kaplan, R. S., "Profit priorities from activity-based costing," *Harvard Business Rev.*, 130–135, May/June 1991.

Crosby, P. B., *Quality Is Free: The Art of Making Quality Certain*, Signet Mentor, New York, 1979.

Deming, W. E., *Out of the Crisis*, MIT Press, Cambridge, MA, 1986.

Ellram, L. M., "Supply chain management: The industrial organization perspective," *Int. J. Physical Distribution Logistics Management* **3**(1), 23–36, 1991.

Garvin, D. A., *Managing Quality: The Strategic and Competitive Edge*, Free Press, New York, 1988.

Gould, L., "How Nike guarantees next day delivery to all its customers," *Modern Materials Handling*, 53–55, September 1992.

Hall, J., "Distribution function is now supply chain integrator," *Ind. Eng.*, 18–19, September 1991.

Hammer, M., "Reengineering work: Don't automate, obliterate," *Harvard Business Rev.* 104–112, July/August, 1990.

Hammer, M., *Beyond Reengineering*. Harper Business, New York, 1996.

Hammer, M., and Champy, J., *Reengineering the Corporation*. Harper Business, New York, 1993.

Imai, M., *KAIZEN: The Key to Japan's Competitive Success*. Random House, New York, 1986.

Kaplan, R. S., and Norton, D. P., *The Balanced Scorecard*. Harvard Business School Press, Cambridge, MA, 1996.

Kjenstad, D., "Coordinated supply chain scheduling," thesis/PhD dissertation, Norwegian University of Science and Technology-NTNU, Department of Production and Quality Engineering, Trondheim, Norway, 1998.

Lalonde, B., and Zinszer, P. H., *Customer Service: Meaning and Measurement*. National Council of Physical Distribution Management, Chicago, 1976.

Lalonde, B., and Cooper, M. C., and Noordewier, T. G., *Customer Service: A Management Perspective*. Council of Logistics Management, Chicago, 1988.

Lee, H., Padmanabhan, P., and Whang, S., "Information distortion in a supply chain: The bullwhip effect," *Management Sci.* **43**(4), 546–558, 1997.

Lyer, A., and Bergen, M. E., "Quick response in manufacturer–retailer channels," *Management Sci.* **43**(4), 559–570, 1997.

Lynch, R. L., and Cross, K. F., *Measure up! Yardsticks for Continuous Improvement*. Blackwell, Oxford, 1991.

Mair, A., *Honda's Global Local Corporation*. St. Martin's, New York, 1993.

Powell, T. C., and Dent-Metcallef, A., "Information technology as competitive advantage: The role of human, business and technology resources," *Strategic Management J.* **18**, 375–405, 1997.

Ravi, K., Stallaert, J., and Whinston, A. B., "Implementing real time supply chain optimization systems," in *Proceedings of the Conference on Supply Chain Management*, University of Hong Kong, Hong Kong, 1995.

Rolstadås, A., "Production planning in the virtual enterprise," in *Proceedings of the First World Congress on Intelligent Manufacturing Processes and Systems*, University of Puerto Rico, 779–789, 1995.

Scott, C., and Westbrook, R., "New strategic tools for supply chain management," *Int. J. Physical Distribution Logistics Management* **21**(1), 23–33, 1991.

Sherman, R. J., "Improving customer service through integrated logistics," *Council of Logistics Management Annual Conference Proceedings*, 1991.

Shigeo, S., *Non-Stock Production: The Shingo System for Continuous Improvement*. Productivity Press, Cambridge, MA, 1988.

Stewart, T. A., "Reengineering: The hot new managing tool," *Fortune,* 41–48, August 28, 1993.

Swaminathan, J. M., "Quantitative analysis of emerging practices in supply chains," PhD thesis, Gradual School of Industrial Administration, Carnegie Mellon University, Pittsburgh, PA, 1996.

Tucker, F. G., Ziyan, S. M., and Camp, R. C., "How to measure yourself against the best," *Harvard Business Rev.* **87**(1), 8–10, 1987.

5

PRODUCT DEVELOPMENT IN A SUPPLY CHAIN

5.1. INTRODUCTION

In a supply chain, the ultimate aim of the five activities of buy, make, move, store, and sell is to deliver a product to the customer. With increasing emphasis on consumer response, and increasing globalization of the market, companies have seen an explosion in the number of different products that they have to offer and an increase in the frequency of new product introduction.

For example, consider the computer industry. The processing power of central processing units (CPUs) doubles every year (this is explained in greater detail in Chapter 6). As a result, chipmakers such as Intel and AMD introduce new chips into the market multiple times per year. Computer manufacturers, such as Compaq, Dell, and Toshiba, in turn launch new computer models that use these faster chips every few months. In addition, these manufacturers supply computers to different parts of the world. Consumer preferences in different markets are different, as are the technical requirements. For example, European countries use a different power supply than does the United States (220 vs 120 V). Also, the French want their interface to the operating system in French, the Germans in German, and the Japanese in Japanese. Undoubtedly, managing a supply chain with these many products is not an easy task.

Established players can no longer afford to concentrate on economies of scale by producing standardized products. If they do, they risk losing the market to niche players that can cater to the needs of a specific market and gain market share at the expense of established players. Thus, companies have to trade off between customer satisfaction and logistical nightmares. In this scenario, new product development and management of the product portfolio are becoming increasingly important strategic factors that influence a company's competitiveness. Thus, product management must address the following interrelated issues:

- What new products should be introduced, at what frequency, and at what time?
- How can the new products be developed in the desired time frame?
- How should the introduction of new products be coordinated with the phasing out of the existing products?

These three issues are not new to the business world. However, new product development has been one of the most closely guarded activities in companies. Product development has primarily been an in-house activity, with selective outsourcing of a few noncritical activities. However, companies are increasingly depending on vendors for components. For example, the automobile manufacturer Daimler-Chrysler outsources more than 70% of the components that go into its automobiles. Therefore, product development must open up to involve and leverage the capabilities of all constituents of the supply chain.

Urban and Hauser (1993, P. 664) outlined the critical success factors for product development during the past three decades as follows:

> The critical success factors that have characterized new-product development have changed in emphasis over time. In the 1970s the five most talked about concepts were (i) market and benefit segmentation, (ii) product positioning and perceptual mapping, (iii) stochastic forecasting models, (iv) creative group methods (e.g., synectics), and (v) idea screening. In the 1980s these concepts became standard components in new-product development, and new ideas and emphasis became prominent. In the 1980s, the critical factors were described as (i) portfolio theory (popularized by the Boston Consulting Group, this portfolio was a matrix of market share and sales growth that defined "cash cows," "stars," "dogs," and "?"), (ii) premarket forecasting and conjoint analysis, (iii) decision support systems and UPC scanner data, (iv) technology/marketing integration and lead users, and (v) competitive strategy and sustainable competitive advantage. In the 1990s the five most popular issues are (i) total quality, (ii) customer satisfaction, (iii) time to market, (iv) manufacturing integration with R&D and marketing, and (v) worldwide strategy and alliances. In the future the most salient concepts will again change.

These emerging concepts are described in the next section.

Due to the evolution of the market, companies have to increase the number of different products they offer to customers and the frequency of new product introductions. Important aspects that companies must focus on are
- Total quality
- Customer satisfaction
- Reduction of time to market
- Manufacturing integration with R&D and marketing
- Worldwide strategy and alliances

5.2. STAGES IN THE PRODUCT LIFE CYCLE

Indeed, we are seeing fresh concepts emerging in new product development. Incorporation of supply chain issues and drawing on the concepts and tools developed in the field of supply chain management will be one of the cornerstones of new product development in the first decade of the 21st century.

With progressively decreasing life cycles of products and the need for the ever-evolving nature of products to adhere to customer demand, the traditional product development activity should be extended to that of product life cycle management. Companies should bring together the planning activities ranging from product conceptualization to phasing out of a product under one umbrella.

The term product life cycle management is defined and interpreted differently. To some, it is the management of issues relating to an individual physical unit from the time of its first manufacturing process to the time of its disposal. To others, it is the management of issues relating to a product, defined as a marketing entity (e.g., Gillete Mach3 razor), from the time it is introduced in the market until it is phased out. This definition of product life cycle has three distinct phases of growth, maturity, and decline (Urban and Hauser, 1993; Polli and Cook, 1969). To us, this represents the sales cycle of the product rather than the life cycle (Fig. 5.1).

The concept of the life cycle of a product beginning with its introduction in the market is relevant for an observer outside the company developing a product. However, for the company itself, the product is "born" much earlier. We define product life cycle management as the set of the following activities:

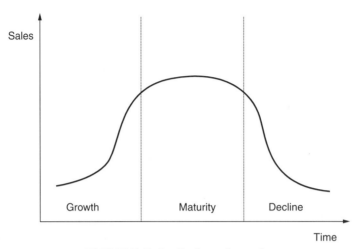

FIGURE 5.1 Product sales cycle.

- Product identification
- Product design and development
- Product introduction
- Product sustenance
- Product phaseout

In this chapter, we briefly describe these activities; however, we concentrate on supply chain issues that should be considered at each stage.

Traditionally, a supply chain is viewed mostly from the operations or logistics perspective. In this chapter, we add the product development organization as an additional constituent of the supply chain. We extend the supply chain concepts to apply them to the product development organization. We also consider the effects of the product development activities on the other parts of the supply chain.

5.2.1. PRODUCT IDENTIFICATION

The first step in the life cycle of a product is the identification of the need for developing a new product. The following are the primary drivers for new product development:

Perceived customer need: The primary sources of this information are the marketing and sales departments of a company. Based on their interaction with customers, they determine what the customers like or dislike about existing products. They explore new features and functionality that customers would like to see in existing or new products. Features could possibly be driven by similar features in comparable competitive products.

Availability of new technology: Availability of new technology can affect product development in two ways: (i) It may enable improvement in the quality and performance of existing products due to better processes and materials or (ii) it may enable entirely new features and products. The first type of change results in incremental product improvement in well-defined and existing market segments. However, the second type of change may create entirely new markets and, hence, is more difficult to manage.

Changes in the environment: A company may be forced to develop new products because of changes in the internal and/or external environment. Changes in the internal environment include the changed financial and strategic goals of the company, such as revenue growth targets, market share, company image, and expansion into new markets. The sales performance of existing products, as indicated by the profit margin and revenue growth, can also affect new product development decisions. If a product or product family are underperforming with diminishing sales and profit margins, they may have to be revitalized with new features and/or replaced by new products. Changes in the external environment include government regulations, actions by competitors, changes in consumer demographics and lifestyles, and changes in other constituents of the supply chain.

A company should take advantage of the expertise of its supply chain constituents at the product identification stage. For example, retailers are the closest to the customers and, hence, have a good feel for changing customer preferences. Also, vendors of components have a view of emerging technologies that the company may use in some of its future products.

Once the need for a product is identified, the company must make a decision whether or not to develop that product. This decision should be based on the answers to two questions:

- How will the development of the new product affect the bottom line of the company?
- Will the development of the new product fit the overall corporate strategy?

Some of the factors affecting the bottom line include the size of the market that the proposed product will target, the amount of investment that would be needed to develop and sustain the product, and the profit margins that can be expected over the life cycle of the product. The company developing the product must also take into account factors such as the presence and strength of competition in the target market, the probability of success of the product, and the magnitude of losses in case of product failure.

Additionally, the company must determine whether the new product fits into the overall corporate strategy: Does the new product utilize the core competencies of the company? How well does it fit with the other products in the company? Is it consistent with the corporate image of the company? How will it affect the other constituents of the supply chain?

If the new product seems promising in one or more aspects discussed previously, then the company must formulate a strategy for product development. The strategy could be innovative or imitative.

In an innovative product development strategy, the company wants to enter the market first. An early market entry is determined to have distinct advantages for a company. Some benefits of bringing products to the market faster than the competition are additional sales revenue, earlier break-even time, extended sales life, premium price, customer loyalty, increased market share, technological lead, innovative image, and increased product range. Robinson (1988) and Robinson and Fornell (1985) provided empirical evidence that companies that enter the market first tend to have substantially higher market share than late entrants. They identified three reasons for higher market share: (i) long-lived marketing mix advantage; (ii) direct cost savings in purchasing, manufacturing, and physical distribution relative to competition; and (iii) relative consumer information advantage. Reaching the market early provides the company with an opportunity to enter the market during the growth phase of the sales cycle. This is the phase in which the profit margins and potential for growth are high and the product has a longer market life.

However, there is significant risk associated with trying to be the first in the market. If it is a new market, then there is little information about the market, which can lead to wrong decisions on the part of the company developing the new product. Not being able to deliver on the promise of being the first in market can have its own risks. Hendricks and Singhal (1997) presented empirical results showing that announcement of delay in product introduction dates leads to a decrease in the market value of firms by approximately 5%. Also, if the innovator does not raise the barriers to entry in the market, imitators can easily follow with competing products and take away market share and profit margins.

Popular product development performance measures of first to market, fast product development, and on-time schedule performance are not the only factors for product success and profitability. Managerial skills, order of market entry, customer understanding, and superior product design are also important factors in the success of a product (Honacher *et al.*, 1987; Lilien and Yoon, 1989; Mitchell, 1991). A company may be better off being an "imitator." It can let a competitor test the new market, learn from its mistakes, and rapidly develop a better product and introduce it to the market.

Christensen (1997) adds an additional dimension of new technology to the product development strategies. He classifies new technologies as sustaining or disruptive. The sustaining technologies are incremental in nature, lead to improved product performance, and cater to established market segments. Disruptive technologies, on the other hand, may result in worse product performance in the short term but bring to the market a different value proposition than established products. Products based on disruptive technologies are typically cheaper, simpler, and easier to use. They underperform established products in mainstream markets. However, they create a market of their own, which in due course starts competing with and then overtakes the established markets. Christensen warns companies against ignoring these disruptive technologies.

The selection of the final product development strategy depends on the corporate culture of the company. However, irrespective of whatever strategy is chosen, before finally committing to development of a new product, the company must review its existing product development projects and the resources committed to them. The company must determine if there are enough resources to develop this new product in a timely manner and, if not, can some existing projects be cannibalized to make resources available for this new project. Of course, any such decision would be based on the relative importance of the two projects in terms of their return on investment and their strategic significance.

5.2.2. PRODUCT DESIGN AND DEVELOPMENT

The design and development phase is often the longest constituent of the time to market of a product. Reducing the product development cycle time can result in a significantly short time to market of the product and, hence, an early market entry for the product. Datar *et al.* (1997) identify three critical considerations in formulating new product development strategies for reducing development time: (i) cross-product learning, (ii) proximity of designers to customers, and (iii) close coordination between designers and process engineers.

There have been several management approaches for reducing product development time. Hauser and Clausing (1988), Griffin and Hauser (1993), and Nayak and Chen (1993) stressed that one should listen to the voice of the customer early in the product development process. Cohen and Levinthal (1990) highlighted the importance of investing in engineering capabilities for rapid learning of processes and translation of customer needs into product design. Clark and Fujimoto (1991) and Iansiti (1993) identified cross-functional teams for concurrent engineering as a tool to reduce product development time. Hayes *et al.* (1988) and Hamel and Prahalad (1991) present the concept of continuous learning through rapid product modifications and line extensions as a means of reducing product development time.

Rauscher (1994) analyzes the development cycle time reduction problem from an organizational perspective. He identifies the following barriers for shortening product development time in an organization:

- Too large a product
- Frequently changing product specifications
- Slow project start-up
- Too many product development projects undertaken by the company at the same time
- Inadequate human resources
- Ignoring queue times during calculations of the lead time
- Inappropriate top management attention
- Lack of concurrency
- Lack of attention to product architecture
- Inappropriate locus of control
- Inappropriate make or buy decisions

The reduction in product development time is also constrained by such factors as learning curves of employees, implementation expenses, and infrastructure restrictions, such as lack of desired manufacturing equipment and capital investments. Wohlin *et al.* (1995) suggested optimizing such factors as human competence, product complexity, and time pressures to reduce product development time. Bayyigit *et al.* (1997) examined how the use of Internet and Intranet in the new product introduction process can reduce product development time by facilitating increased levels of communication within a company and between a company and its customers.

Cohen *et al.* (1996) present a multistage model of a new product development process to determine the trade-off between product development cycle time and improvements in product performance. They show that if product improvements are additive (over stages), it is optimal to allocate maximal time to the most productive development stage. They also show that factors such as the size of the potential market, the presence of existing and new products, profit margins, the length of the window of opportunity, the firm's speed of product improvement, and competitor product performance affect the time-to-market and product performance targets.

The product development team should also take into account the availability of product development resources to a new project and the requirements that the new product would place on these resources. This would affect the time it takes to design and develop the product. An approach to maximize the utilization of development resources by involving them in multiple projects at the same time results in each project getting only a fraction of their total capacity. This may result in all the projects being prolonged. Although dedication of resources to a single project may not be the prudent option, organizations should try to limit the number of projects that resources are working on concurrently. This will help them to concentrate more effectively on the projects at hand and complete them quicker. Thus, if instead of working on four projects simultaneously and finishing them in 1 year, a team works on one project at a time and finishes each in 3 months, the productivity is similar. However, the work-in-process of development tasks as well as the development time are reduced significantly.

It is cautioned that any attempt to reduce the development time of a product will have a cost associated with it. Sometimes the cost may exceed the possible gains from a shortened product development cycle. Smith and Reinertsen (1998) presented simple tools for quick what-if analysis to determine whether or not the cost of an action to reduce the development time is acceptable.

We divide the design and development phase into the following stages:

- Translating customer requirements into product specifications
- Conceptual product design and product architecture
- Detailed design
- Prototyping and testing

5.2.2.1. Translating Customer Requirements into Product Specifications

Irrespective of whether the driver for new product development is perceived customer need, availability of a new technology, or change in the environment, the ultimate aim of the product is to satisfy customer requirements. Therefore, a company must determine what specific needs of the customer the new product will satisfy. In the case of the latter two drivers, the company may have to spend significant effort to determine the customer requirements to which the product would cater. The product development team then needs to translate the customer requirements into the functional specifications for the product. Note that these are not the technical specifications of the product, which are developed at the conceptual design and product architecture stage. The tools and techniques needed for translating customer requirements into functional specifications are explained in detail by Urban and Hauser (1993) as well as Smith and Reinertsen (1998) and include such popular techniques as quality function deployment (QFD). Input from other constituents of the supply chain, such as retailers and aftersales service providers, should be obtained in this process.

5.2.2.2. Conceptual Design and Product Architecture

Arguably, this stage has the greatest impact on the supply chain of the product and can provide the greatest benefit to the organization by incorporating supply chain concerns into product design.

Conceptual Design

Conceptual design of the product includes the task of translating the functional specifications of a product into technical specifications. This is the stage in which the product development team determines what physical entities in the product will satisfy the various functional requirements and also the available technologies that will enable the realization of these physical entities. The product development team analyzes and then selects the technologies that will be used in the different components of the product. These technologies include the physical principles on which the design of the different components will be based as well as the materials and manufacturing processes that will be used to manufacture the different components. The output of this stage is the overall architecture of the product, the preliminary selection of manufacturing processes for making the various components, and the detailed technical specifications of the product components.

In making these decisions, the product development team should collaborate with other constituents of the supply chain, including the external suppliers of raw materials and components as well as the internal departments, such as research and engineering, manufacturing, operations, and logistics.

The external vendors can provide information regarding the latest technology that they are using as well as their future plans for the development of new technology. The product development team may collaborate with vendors on the development of components that they will supply. Involving them at this stage makes the design of the interfaces to these supplied components much less prone to errors.

The research and engineering department can provide information regarding the status of the different technologies being developed. This information should be used in determining what technologies can be used in the new product, based on the status of their development and testing. The manufacturing department can provide information regarding the capabilities of the different manufacturing processes within the company. The limitations of the manufacturing resource should be taken into account while designing the product. The operations and logistics departments can provide information regarding the availability of production, inventory, and transportation resources that would be needed by the new product when it enters the production phase. These issues should be taken into consideration in product design.

Govil (1999) presented a new design methodology for incorporating production concerns at the conceptual design stage. He noted that if the manufacturing processes selected for the production of a new product are expected to be heavily utilized by other existing products at the desired launch time, then the production system may not be able to meet the demand of the new product. If demand exceeds supply by a significant amount during product launch, the consequence can be severe in terms of costs, including direct expenses and jeopardized investment in the form of order cancellation, poorly timed and wasted promotional effort, and, in extreme cases, total failure of the project with R&D and capital equipment investment losses.

Govil presented techniques to take resource capacities at the time of product launch into account during the conceptual product design stage. Options include the following:

- Use of alternate manufacturing processes and materials that will avoid the use of bottleneck resources without significantly affecting the cost and functionality of the product
- Modifications to the attributes of the product components to reduce their processing time on critical resources
- Outsourcing of certain components that require the bottleneck resources

These steps can enable the company to attain the desired production volume at product launch. If technically the selection of alternate processes or modifications to component attributes is not feasible, then the product development team should plan to initiate the acquisition and installation of additional resources in synchronization with the product development activities. The techniques presented by Govil (1999) can enable the product development team to determine what combination of component attribute modification, alternate manufacturing process selection, and increased resource capacity will provide the maximum increase in production with minimum investment.

Product Architecture

Product architecture significantly affects the entire supply chain in terms of the cost of operation and efficiency. Decisions made at the product architecture stage include the following:

- The selection of the technology to be used
- The selection of the components to manufacture in-house and those to outsource
- The sequence in which the components are to be put together to complete the products

The more modular the product, the more subsystems can be designed in parallel, reducing the product development cycle time. However, this may result in additional product cost due to redundancy in the different modules as well as complications in the interfaces between the different modules.

Worldwide competition leads to product proliferation and demand uncertainty, which makes it impossible to obtain precise, and thus useful, forecasts. In turn, inaccurate forecasts require high inventory investments and lead to poor customer service.

Most enterprises strive for strategies that allow companies to respond quickly to various customer requirements (reactivity) while keeping inventories and work-in-process at reasonable levels. To reach this goal, they redesign processes and/or products to defer the point of differentiation (PD) downstream — that is, as close as possible to the final production stage. As a consequence, they can manufacture until the PD based on global forecast, which is quite precise due to the number of semifinished products to manufacture, and proceed on a make-to-order basis from the manufacturing stage that follows the PD onwards. The closer the PD to the completion stage, the less amount of time required to deliver the product, and thus the more reactive the supply chain. Usually, however the closer the PD to the completion stage, the lower the number of different products that can be derived from the semifinished products obtained at the PD level, as shown in Fig. 5.2.

5.2. STAGES IN THE PRODUCT LIFE CYCLE

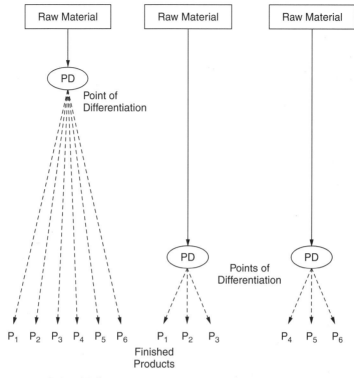

FIGURE 5.2 PD location vs. product type spectrum.

This introduces a risk due to the versatility of the demand. This may also pose a problem for the forecasts since the reduction in the number of different products that can be derived from the semifinished products at the PD level usually implies a reduction in the volume of semifinished products to be forecasted, and thus results in less accurate forecasts. The responsibility of supply chain managers is to find the best trade-off between reactivity on one hand and inventory costs as well as risk of inaccurate production on the other hand.

Commonality is another way to reduce production costs. Commonality is a strategy that consists of developing components that are part of a large number of different products. Often, manufacturing processes require different components or semifinished products at different stages for assembly operations. Increasing the level of commonality at an early stage of the manufacturing process may delay the PD without increasing the risk associated with the inventories at the PD level. The explanation is straightforward: A high level of commonality reduces the number of possible combinations of the components or semifinished products (also called modules) and thus pushes naturally downstream the PD of products. Commonality also helps in designing products in a more modular fashion.

Cost Analysis of Deferment Design

Redesigning products in order to defer the PD requires extra investments, although redesign will provide some additional benefits. A precise evaluation of costs and benefits should be conducted before deciding whether reengineering is necessary. Factors that should be considered in the evaluation include the following:

- Salaries of the designers
- Costs of the new resources required by the new design
- Cost for training employees
- Cost for changing the layout of the manufacturing system
- Cost for advertising since customers should be kept informed of the efforts made to increase the quality and the effectiveness of the products as well as the quality of the service (reactivity)

These costs, except for salaries of the designers, are difficult to evaluate since they will really be known only at the end of the redesign process.

Benefits are even more difficult to evaluate. The most important factors that generate benefits are as follows:

- The reduction of inventory and handling costs: Inventory costs are reduced due to the higher level of commonality that increases the volume of semifinished products and components to be managed, which allows a global reduction of the inventory for a constant service level.
- The reactivity of the system, which increases its competitiveness and helps win new customers and markets.

Production System Design

In addition to conceptual product design and product architecture, the design of the production system that will manufacture the new product should also be initiated at this stage. In many cases, existing production systems will be used, with some changes to processes or the addition of some new resources. The issue of whether to use existing manufacturing processes or use/develop new ones is important. The use of existing processes reduces the amount of risk associated with the manufacturing of the new components. It also saves time with regard to acquiring new resources, testing the new processes, and training the operators regarding the resources involved with the new processes. However, the performance of new products should not be compromised for these savings unless the use of new processes is economically or technically not viable. Smith and Reintersen (1998) argue that, if possible and economically viable, new processes should be tested on some of the existing products so that they are available in a timely manner for use in producing new products.

At this stage the product development team should consult the suppliers of any new resources. Their input can be useful in determining the technical specifications for components that are affected by specifications of the new resources.

The concurrent design of the production system results in a significant reduction in the time to market of the new product because acquisition, installation, and testing of new resources is a lengthy process.

5.2.2.3. Detailed Design

The conceptual design of the product in terms of the product architecture, the selection of technologies and processes for manufacturing of the different product components, and the detailed specification for each product component are the input, to the detailed design stage.

The detailed design should incorporate additional downstream issues in the product life cycle into the product design. In the 1980s and 1990s, many techniques and methodologies were developed to attain this goal. These techniques and methodologies are commonly known as design for "X" (DFX), where X is the parameter that the methodology proposes to incorporate into product design.

The first of these DFXs was design for manufacture (DFM) (Bralla, 1986). The guiding principle in DFM is to design products such that they are easy to manufacture. The designers specifically take into account the limitations of manufacturing processes that will be used to make the product components. For example, if turning is to be the manufacturing process, then the part can only have features with circular symmetry. Similarly, for parts that are to be made by casting, one has to consider the thickness of the different sections and the change in thickness from one section to another. Abrupt changes in section thickness result in product defects due to uneven solidification of the molten metal. Incorporation of manufacturing consideration in product design can reduce the manufacturing time of components. This also prevents any redesigns at late stages in product development, which can be very costly and time-consuming. Errors in product design due to lack of consideration of processing issues can possibly result in defective components, some of which may be detected at the testing stage and end up as scrap, whereas others may slip through the testing stage, resulting in many products being returned by customers. Redesign at late stages is very expensive. Thus, DFM is a powerful tool and is becoming a standard practice in industry.

The next most influential methodology is design for assembly (DFA) (Boothroyd *et al.*, 1994). DFA emphasizes the importance of incorporating assembly issues into product design. These considerations include the ease of assembly (thereby reducing assembly time) as well as analysis of product architecture with an aim of identifying components that could be combined. The driving philosophy behind the methodology is to minimize the assembly time by improving the design of interfaces as well as to minimize the number of components in the assembly. DFM and DFA merged into design for manufacture and assembly (Boothroyd, 1994; Bryant *et al.*, 1994).

These methodologies were soon followed by methodologies that incorporated other downstream issues such as design for disassembly, reliability, maintainability, environment, ergonomics, aesthetics, schedulability, and synchronized flow manufacturing (Foo and Lien, 1995; Moss, 1985; Blanchard *et al.*, 1995; Fiksel, 1996; Kusiak and He, 1994; Ahmadi and Wurgaft, 1994).

The availability of computer-aided design (CAD) packages significantly improved productivity in detailed design. The CAD packages of the 1990s have been much improved from the packages of the 1970s and 1980s, which were little more than electronic drafting boards. The latest state of the art in CAD packages offers the ability to render parts in three-dimensions. This representation is often helpful in identifying many design problems. These packages also allow virtual assembly of components. Thus, different groups do not have to wait until the prototype stage for their components to be tested for fit with one another, thereby significantly reducing redesign efforts and time. Some of the packages also incorporate some DFX techniques.

5.2.2.4. Prototyping and Testing

Prototyping and testing have been revolutionized by new technology. Although the testing of physical prototypes has not been eliminated, its use has been minimized by the virtual testing environments offered by CAD packages. One can not only visualize the components in three-dimensions, but also assemble components, and perform tests such as stress analysis, flow analysis, and thermal analysis. Thus, components being developed by different departments and even outside vendors can be made available electronically to the other members of the product development team. This helps in identifying potential problems early.

The availability of a new technique called free-form fabrication has significantly reduced the product prototyping task. This technique is also called rapid prototyping. In this technique, special machines create test components with complicated geometrical shapes without using tools or dies. The input to these machines is the three-dimensional CAD drawing of the object. Based on this input, the rapid prototyping machines create a solid of the input shape using materials such as polymers, paper, and powdered metals. These models can be used for visual verification or as molds and dies for creating parts or even actual parts. However, as per the current state of the art in this field, these rapid prototyping machines cannot be used for mass production because of their high cost and slow speed of manufacturing. Also, they can handle only certain materials. Some of the more popular rapid prototyping techniques are stereolithography (Jacobs, 1992), selective laser sintering, and fused deposition modeling. Kochan (1993) provided a comprehensive list of the various techniques for free-form manufacturing.

These new testing and prototyping techniques can significantly reduce investment in new resources and tools that may never be used if the product is not mass produced. Also, each new iteration in the design often requires new tools, which can be avoided by these techniques. The time saving achieved also reduces the time to market of the product.

Instead of testing the completely assembled product, testing of individual components should be done as soon as their design is complete. This exposes the functional defects in the components earlier and, hence, they can be addressed at relatively lower cost and early in the development cycle. Similarly, testing of various subassemblies should be carried out as soon as the design of all their constituents is complete. The goal of testing the complete product should only be to test the interfaces between the different subassemblies. As testing of complete products is expensive, any failures due to component defects at this stage should be avoided by prior component testing.

In addition to testing of the physical and functional attributes of the product, some companies test perceptual characteristics of the product, including customer liking, ease of use, and image perception. Different companies may employ a different level of product testing, ranging from evaluation of product by experts to evaluation of product by a select group that is representative of the total market, testing of product in a small geographical area, or large-scale test marketing in a major market (a few large cities). Many considerations go into the level of test marketing that is done. Test marketing reduces the risk of product failure during full-scale launch. On the other hand, a large-scale test marketing effort may result in delayed product introduction, significant costs involved in the logistics of the test marketing effort, and significant investment in materials and equipment because a large-scale test marketing effort involves product quantities that can be produced only by mass production equipment. Large-scale test marketing may also result in the quick introduction of an imitative competitive product. These factors should be taken into account when deciding on the level of test marketing.

5.2.3. PRODUCT INTRODUCTION

Of all the stages in the life cycle of a product, its launch or introduction requires the largest commitment of time, money, and management resources (Urban and Hauser, 1993; Hultink *et al.*, 1997). Product introduction includes those decisions and activities that are necessary to present a product to its target market and begin generating income from sales of the new product. Product introduction is often considered the most expensive, most risky, and least managed part of the overall product development process (Calantone and Montoya-Weiss, 1993). Hultink *et al.* (1997) discuss five key issues in product launch:

- What to launch
- Where to launch
- When to launch
- Why to launch
- How to launch

They state that these launch decisions have a major impact on the success of a new product. Of the five issues, the first four (namely, what, where, when, and why to launch) are strategic decisions and are made long before the actual introduction of the product. In our scheme, these issues are addressed at the product identification stage. How to launch a product is a tactical decision. However, strategic factors have a significant impact on the tactical decision.

One of the approaches in product launch is to go for a "big bang" with the introduction of the product in all markets simultaneously. The second approach is a "roll-out" approach in which the product is launched in only limited regions at first and is then gradually rolled out to other regions.

5.2. Stages in the Product Life Cycle

There are positives and negatives for both approaches. A gradual roll out requires a significantly less amount of logistical effort and investment and is also less risky. However, this may present opportunities for competitors that may enter the markets that are left open. A big-bang approach avoids this vulnerability but with the added cost of significant risk and the requirement of a monumental logistical and marketing effort.

From a marketing perspective, the issues in product launch involve the pricing of the product, the type and intensity of advertising and the promotional effort that should accompany the product launch, the variations of the products that should be introduced in the different markets, and the exact timing of the launch. These decisions should be made in conjunction with input from the logistics group, the sales group, and other constituents of the supply chain, such as suppliers and retailers.

The level of advertising and promotions should be in line with the expected feasible production rate of the new product. If the demand created by marketing cannot be satisfied by timely production, it results in unsatisfied customers and opens doors for competitors to tap this vulnerable market. Thus, production start-up and ramp-up should be in line with the marketing strategy and vice versa.

Marketing decisions should be conveyed upstream to the suppliers of raw materials and components so that they can plan their production to meet the expected demand. If suppliers cannot meet the expected demand, this information should be passed on to production and marketing so that they can modify their strategy.

Similarly, marketing and production decisions should be conveyed downstream to the retailers. This enables them to create the appropriate stocking and display infrastructure for the new product. They may also have to set a pricing strategy for the new product and update the different information systems (e.g., bar code scanners, product and inventory databases, and receiving and marking systems) and business practices involved with the retailing of this new product. This may involve training of employees at the retailer. A delay on the part of the retailer to do so may result in a delay in product introduction and poor customer service.

The sales and service organizations should also be an integral part of product introduction. Timely training of the sales representatives may become a crucial factor in the success of the product. An untrained sales representative may make the customer uncomfortable and reluctant to buy the new product. An effective incentive plan for sales representatives is needed for them to give the new product the extra push that it needs to become established. During the product introduction phase, good customer service is needed so that customers gain confidence in the new product. Thus, successful product introduction requires effective coordination between all the constituents of the supply chain.

5.2.4. PRODUCT SUSTENANCE

After the successful introduction of a product, product sustenance is the phase in which a company attempts to make the product available to the right people, at the right time, at the right price. The product sustenance phase is dominated by logistics and marketing groups. The logistics groups use traditional supply chain management techniques of demand management, supply management, production planning, inventory management, transportation planning, demand fulfillment, and others to maximize the efficiency of the organization. The marketing groups try to create a positive image for the product to gain and sustain market share.

Cooperation among all the constituents of the supply chain is essential in obtaining operational efficiency in this phase. Also of importance is the after sales service offered for the product. Most durable goods require some repair during their lifetime. Although high product quality can minimize the requirement for repairs, they cannot be eliminated. Therefore, manufacturers must ensure that efficient mechanisms for a high service level are in place when a product is brought to the market. A key concern is that the suppliers of critical components should not go out of business or stop production of the components before the product is phased out. This should be one of the criteria for initial selection of vendors. Any such occurrence can severely affect sales in the most revenue-intensive phase of the product.

This is the phase that brings in most of the revenues from the product. However, many companies become so involved in maximizing the revenues and profits from the products that they miss an essential point: *This is only one of the phases in the life cycle of the product and must come to an end.* Therefore, many companies fail immediately after their best years. The product sustenance phase should be the incubator for the next generation of products.

There are a multitude of indications regarding what the next product should be. First are the customers and the existing product. An analysis of product returns provides one of the best sources of new improvements. Products can be returned for a variety of reasons:

Product has manufacturing defects: This indicates a deficiency in the manufacturing processes or in the design of the product.

Therefore, product design must be re-examined and, if needed, the manufacturing processes must be improved for use in existing and new products.

Product is difficult to use: This indicates that customer requirements were not properly determined or were not properly incorporated into product design. The product should be reevaluated against these customer requirements and redesigned as an improvement to the existing product or as a new product.

Product performance does not live up to the expectations: This indicates that either the management of customer expectations was not handled properly by the marketing and salespeople or the product design could not meet customer requirements. In any case, the product development team, including marketing and design people, should reexamine these issues.

A comparable product is available at a better price: This indicates that the competitor is achieving efficiency in operations because of some factors. These could be more favorable pricing from the vendors, use of less expensive raw materials, more efficient manufacturing processes, or more efficient logistics. According to the reason identified, the necessary remedial action should be taken. If it is not possible to compete at the competitor's price, the product management team should explore the option of differentiating the product from its competitor's based on the functionality offered, the quality of the product, or the image of the product. These issues should then be addressed in the development of new products.

The second important source of new product ideas is the competition. Once a company establishes or creates a market for a product, it is followed by several competitors that also want to cash in on that market. To compete, market leaders have to stay ahead of competitors in innovation or at least offer products that have similar features. Many established companies are good at this. However, they sometimes ignore products that do not directly challenge their products. Often, these products are based on new technology and are cheaper, inferior, and cater to a specific and small market segment. Many of these products fail or remain confined to a small market. Sometimes, however, the new technology improves and establishes a large new market, often at the expense of existing products. Christensen (1997) provides numerous examples of this phenomenon. Complacency in following and reacting to new technology can and often does lead to failure of established market players.

5.2.5. PRODUCT PHASEOUT

Product phaseout is the final stage in a product's life cycle. However, its effect on other phases of the product life cycle and the supply chain seldom receives the recognition that it deserves.

A product may need to be phased out for many reasons, including the following:

- *The product technology becomes outdated:* Currently in the computer industry, the most commonly used processors have speeds in excess of 400 MHz. Thus, it is difficult for a company to sell computers with 100-MHz processors even if they are slightly cheaper.
- *Customer preferences change:* The apparel industry has to constantly keep abreast of the latest trends in fashion.
- *Supply of raw materials or outsourced components becomes unavailable or prohibitively expensive:* In this age of touch-tone phones, it is becoming increasingly difficult and expensive to make and sell rotary phones.
- *Cheaper alternatives become available:* The television industry has seen continually decreasing prices. As newer and cheaper products entered the market, the older, more expensive products were forced out.
- *A new product is entering the same market as the old one:* Microsoft introduced its new PC operating system Windows 98 to replace Windows 95.
- *The product is coming to the end of its intended life cycle:* Automobile manufacturers introduce a new model every year. Thus, every model is intended to be in the market only for 1 year and is phased out as the new model comes in.

A telltale sign that a product has become a candidate for phaseout is steadily decreasing sales and profit margins. However, except for a few obvious cases (yearly automobile models), it is difficult to determine if the product has reached that stage. It could possibly be due to cyclical or seasonal sales, or maybe the product can be revived by proper marketing efforts. Adding some new features to a product that make a more attractive or useful may possibly increase the product life. The company may have to persist in the market even at reduced margins for strategic reasons, including corporate image, product breadth, and market share. Only after a thorough analysis that incorporates these considerations has been carried out should a decision regarding the phase out of a product be made.

5.2. STAGES IN THE PRODUCT LIFE CYCLE

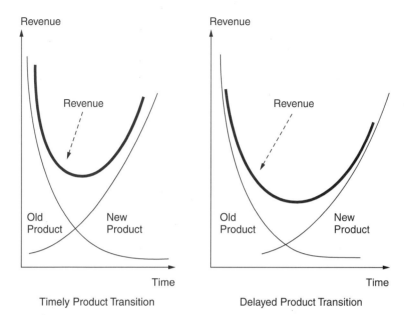

FIGURE 5.3 Difference in revenue due to product transition timing.

A proactive product management team would already have a new product in the pipeline to replace the aging product. The question is when and how to phase out the product and replace it with the new product. The timing of the phaseout can have a significant impact on revenues and profit margins. If the product is phased out too early, then potential revenue from it is lost. If it is phased out too late, the potential from the sale of a higher margin new product is lost. Thus, a company has to determine the optimal time of phasing out a product (Fig. 5.3) by analyzing the cash flows from different timings of old product phaseout/new product launch.

One has to be very careful in computing the cost of products with decreasing sales volume. Accounting practices based on standard costs may underestimate the cost of products. Decreasing sales volume increases the cost of the product because the fixed overhead is distributed among fewer units.

Similarly, maintaining high customer service levels for these products becomes more expensive. It is cautioned that in the case of durable goods, even after the phasing out of the product, service parts should be available for a reasonable time to avoid unhappy customers.

Phasing out a product can also present unique opportunities for a company. If a company determines that a product must be phased out, it can significantly decrease advertising and promotions for that product (or eliminate them) and thereby save a significant amount of resources and money.

The same resources can then be used for launching a new product. Phasing out of a product also frees up production capacity on manufacturing resources. This capacity can then be used for new products. Thus, proper coordination between the phasing out of an old product and the introduction of a new product can result in optimal use of the production resources.

Phaseout decisions should be conveyed upstream to the suppliers so that they can plan their production accordingly. Similarly, they should be conveyed downstream to the retailers so that they can plan how best to handle this through efforts such as clearance sales. One should be careful so as not to affect the sale of the new product due to discounts on the old products. The two promotional efforts should complement each other and not compete with each other.

Product life cycle management comprises the following steps:
- Product identification
- Product design and development
- Product introduction
- Product sustenance
- Product phaseout

This chapter addressed these steps and showed how they could by improved by the supply chain philosophy.

5.3. CONCLUSION

Products play a pivotal part in supply chains. The goal of this chapter was to show how the life cycle of products can be easily integrated in the supply chain framework. We also showed how it is possible to further improve supply chain efficiency by acting at the different stages of the life cycle of the products.

REFERENCES

Ahmadi, R. H., and Wurgaft, H., "Design for synchronized flow manufacturing," *Management Sci.,* **40**(11), 1469–1483, 1994.

Aviv, Y., and Federgruen, A., "The benefits of design for postponement," in *Quantitative Models for Supply Chain Management* (S. Tayur, R. Ganeshan, and M. Magazine, Eds.), pp. 553–584, Kluwer, Boston, 1998.

References

Baker, K. R., "Safety stocks and component commonality," *J. Operations Management* **6**(1), 13–22, 1985.

Ballon, R. H., *Business Logistics Management,* Prentice Hall, New York, 1992.

Bayyigit, A. C., Inman, O. L., and Kuran, E. D., "New product introduction: Reducing time to market using Internet and Intranet technology," in *Innovation in Technology Management. The key to Global Leadership. PICMET '97 Portland International Conference on Management of Engineering Technology,* Portland State University, pp. 454–459, 1997.

Blanchard, B. S., Verma, D., and Peterson. E. L., *Maintainability: A Key to Effective Serviceability and Maintenance Management* Wiley, New York, 1995.

Boothroyd, G., "Product design for manufacture and assembly," *Computer Aided Design,* **26**(7), 505–519, 1994.

Boothroyd, G., Dewherst, P., and Knight, W., *Product Design for Manufacturing and Assembly.* Dekker, New York, 1994.

Bralla, J. G., *Handbook of Product Design for Manufacture.* McGraw-Hill, New York, 1986.

Bryant, R. V. E., Laliberty, T. J., and Lapointe, L. J., "DICE MO — A collaborative DFMA analysis tool," *IEEE Trans. Eng. Management,* 96–103, 1994.

Burhanuddin, S., and Randhawa, S. U., "A framework for integrating manufacturing process design and analysis," *Computers Ind. Eng.* **23**(1–4), 27–30, 1992.

Calantone, R. G., and Montoya-Weiss, M. M., "Product launch and follow on," in *Managing New Technology Development* (W. E. Souder and J. D. Sherman, Eds.), McGraw-Hill, New York, 217–248, 1993.

Chen, F., Ryan, J., and Simchi-Leiv, "The impact of exponential smoothing forecasts on the bullwhip effect," Working Paper, Northwestern University, 1997.

Child, P., Diederichs, R., Sanders, F., and Wisniowski, S., "The management of complexity," *Sloan Management Rev.,* 73–80, Fall 1991.

Christensen, C. M., *The Innovator's Dilemma: When New Technologies Cause Great Firms to Fail.* Harvard Business School Press, Boston, 1997.

Christopher, M., *Logistics and Supply Chain Management: Strategies for Reducing Costs and Improving Services.* Financial Times/Pitman, London, 1992.

Christopher, M., *Logistics and Supply Chain Management.* Financial Times/Irwin, Burr Ridge, IL, 1994.

Clark, K. B., and Fujimoto, T., *Product Development Performance: Strategy, Organization and Management in the World Auto Industry.* Harvard Business School Press, Boston, 1991.

Cohen, M. A., Eliashberg, J., and Ho, T. H., "New product development: The performance and time-to-market tradeoff," *Management Sci.* **42**(2), 173–186, 1996.

Cohen, W. M., and Levinthal, D. A., "Absorptive capacity: A new prescriptive to learning and innovation," *Administration Sci. Q.* **35**, 1990.

Collier, D. A., "Aggregate safety stock levels and component part commonality," *Management Sci.* **28**(11), 1296–1303, 1982.

Copacino, W. C., *Supply Chain Management: The Basics and Beyond.* St. Lucie Press, Falls Church, VA, 1997.

Datar, S., Jordan, C., Kekre, S., Rajiv, S., and Srinivasan, K., "New product development structures and time-to-market," *Management Sci.* **43**(4), 452–464, 1997.

Dodd, T. H., Pinkleton, B. E., and Gustafson, A. W., "External information sources of product enthusiasts: Differences between variety seekers, variety neutrals, and variety avoiders," *Psychol. Marketing* **13**(3), 291–304, 1996.

Fiksel, J. (Ed.), *Design for Environment*. McGraw-Hill, New York, 1996.

Foo, S. W., and Lien, W. L., "Reliability by design: A tool to reduce time-to-market," *IEEE 1995 Eng. Management Conf.,* 251–256, 1995.

Garg, A., and Tang, C. S., "On postponement strategies for product families with multiple points of differentiation," *IIE Trans.* **29**, 641–650, 1997.

Govil, M., *"Integrating product design and production: Designing for time-to-market,"* PhD dissertation, Department of Mechanical Engineering, University of Maryland, College Park, 1999.

Griffin, A., and Hauser, J. R., "The voice of customer," *Marketing Sci.* **12**, 1–27, 1993.

Hamel, G., and Prahalad, C. K., "Corporate imagination and expeditionary marketing." *Harvard Business Rev.,* 81–92, JulyAugust 1991.

Hammer, M., and Champy, J., *Reengineering the Corporation.* Harper, New York, 1993.

Hauser, J. R., and Clausing, D., "The house of quality," *Harvard Business Rev.,* **66**(3), 63–73, 1988.

Hayes, R. H., and Wheelright, S. C., *Restoring Our Competitive Edge.* Wiley, New York, 1984.

Hayes, R. H., Wheelright, S. C., and Clark, K. B., *Dynamic Manufacturing: Creating the Learning Organization.* Free Press, New York, 1988.

He, D. W., and Kusiak, A., "Design of assembly systems for modular products," *IEEE Trans. Robotics Automation* **13**(5), 646–655, 1997.

Hendricks, K. B., and Singhal, V. R., "Delays in new product introductions and the market value of the firm: The consequences of being late to market," *Management Sci.* **43**(4), 422–436, 1997.

Houlihan, J., *Exploiting the Industrial Supply Chain, Manufacturing Issues.* Booz Allen & Hamilton, Mclean, VA, 1987.

Honacker, V., Wilfried, R., and Day, D., "Cross-sectional estimation in marketing: Direct versus reverse regression," *Marketing Sci.* **6**, 254–267, 1987.

Hultink, E. J., Griffin, A., Hart, S., and Robben, S. J., "Industrial new product launch strategies and product performance," *J. Product Innovation Management,* **14**, 243–257, 1997.

Iansiti, M., "Real world R&D: Jumping the product generation gap," *Harvard Business Rev.,* 138–147, MayJune 1993.

Jacobs, P. F., *Rapid Prototyping and Manufacturing: Fundamentals of Stereolithography.* SME, Deaborn, MI, 1992.

Kochan, D., *Solid Freeform Manufacturing: Advanced Rapid Prototyping.* Elseveir, New York, 40–53, 1993.

Kusiak, A., and He, W., "Design of components for schedulability," *Eur. J. Operational Res.* **76**, 49–54, 1994.

Lambert, D., and Slater, S., First, fast, and on-time. The path to success. Or is it? *Managing Virtual Enterprises IMEC 96,* 275–280, 1996.

Lee, H., "Effective management of inventory and service through product and process redesign," *Operations Res.* **44**, 151–159, 1996.

Lee, H., Billington, C., and Carter, B., "Hewlett-Packard gains control of inventory and service through design for localization," *Interfaces* **23**, 1–11, 1993.

Lee, H., and Tang, C. S., "Modeling the costs and benefits of delayed product differentiation," *Management Sci.* **43**, 1997.

Lee, H., and Tang, C. S., "Variablility reduction through operation reversal," *Management Sci.* **44**, 162–173, 1998.

References

Lilien, G. L., and Yoon, E., "Determinants of new industrial product performance: A strategic reexamination of the empirical literature," *IEEE Trans. Eng. Management,* **36**(1), 3–10, 1989.

Macduffie, J. P., Sethuraman, K., and Fisher, M. L., "Product variety and manufacturing performance: Evidence from the international automotive assembly plant study," *Management Sci.* **42**(3), 350–369, 1996.

Mitchell, W., "Dual clocks: Entry order influences on incumbent and newcomer market share and survival when specialized assets retain their value," *Strategic Management J.* **11**, 85–100, 1991.

Moss, M. A., *Designing for Minimal Maintenance Expense.* Dekker, New York, 1985.

Nayak, P. R., and Chen, A. C., "Listening to customer," in *Prism.* Little, Cambridge, MA, 1993.

Nevins, J. L., and Whitney, D. E., *Concurrent Design of Product and Processes.* McGraw-Hill, New York, 1989.

Pawar, K. S., Menon, U., and Riedel, J. C. K. H., "Time to market," *Integrated Manufacturing Systems,* **5**(1), 14–22, 1994.

Polli, R., and Cook, V., "Validity of product life cycle," *J. Business* **42**(4), 385–400, 1969.

Rauscher, T. G., "Time to market problems — The organization is the real cause," *IEEE Trans. Eng. Management,* 338–345, 1994.

Roberts J., "Formulating and implementing a global logistics strategy," *Int. J. Logistics Management,* **1**(2), 53–58, 1990.

Robinson, W. T., "Sources of market pioneering advantages: The case of industrial goods industries," *J. Marketing Res.* **25**, 87–94, 1988.

Robinson, W. T., and Fornell, C., "Sources of market pioneering advantages in consumer goods industries," *J. Marketing Res.* **22**, 305–317, 1985.

Ross, D. F., *Competing through Supply Chain Management: Creating Market-Winning Strategies through Supply Chain Partnerships.* Chapman & Hall, New York, 1998.

Roy, R., and Pofter, S., "Managing engineering design in complex supply chains," *Int. J. Technol. Management* **12**, 403–420, 1996.

Scott, C., and Westbrook, R., "New strategic tools for supply chain management," *Int. J. Phys. Distribution Logistics Management* **21**(1), 23–33, 1991.

Smith, P. G., and Reinertsen, D. G., *Developing New Products in Half the Time: New Rules, New Tools.* Wiley, New York, 1998.

Sterman, J. D., "Modeling managerial behavior: Misperceptions of feedback in a dynamic decision making experiment," *Management Sci.* **35**, 321–339, 1989.

Underhill, T., *Strategic Alliances: Managing the Supply Chain.* PennWell, Tulsa, OK, 1996.

Urban, G. L., and Hauser, J. R., *Design and Marketing of New Products.* Prentice Hall, Englewood Cliffs, NJ, 1993.

Wang, C. H., and Sturges, R. H., "Concurrent product/process design with multiple representation of parts," *IEEE Int. Conf. on Robotics & Automation* **3**, 298–304, 1993.

Wohlin, C., Xie, M., and Ahlgren, M., "Reducing time to market through optimization with respect to soft factors," *Proc. IEEE Eng. Management Conference,* 116–121, 1995.

6

ENABLING TECHNOLOGIES

Computer technologies have been one of the major enablers of the current supply chain planning paradigms. This chapter presents a brief review of the current state of the art in these technologies as well as their development over time. The chapter is divided into two parts. The first part discusses the developments and the current state of the art in the field of computer technology. The second part focuses on computer applications that support the supply chain planning paradigms and practices.

6.1. TECHNICAL ENABLERS: DEVELOPMENTS AND THE STATE OF THE ART IN COMPUTER TECHNOLOGY

The use of computers in businesses dates back to the 1960s. However, their widespread use was severely restricted by the capabilities as well as the cost of the hardware. Also, human resources trained in using these "highly sophisticated systems" were required. The progress of computers from those days to the current environment in which computers have proliferated most areas of business can be attributed to four main developments:

- Hardware development
- Software development
- Business needs
- Human resource development

6.1.1. HARDWARE DEVELOPMENT

6.1.1.1. Computer Hardware

Computer hardware has a come a long way from the gigantic mainframe machines of the 1960s that occupied entire rooms to the powerful desktop and laptop PCs of the present. On the same note, personal digital assistants (PDAs) of today are more powerful than the fastest desktop PC of the early 1980s. Progress has not just been in reduction of size and cost but also in increase of power of both the microprocessor (CPU) and the main memory (RAM). This phenomenon is captured by the predictions of two of the pioneers in the area of computer hardware. In 1964 Intel founder Gordon Moore predicted that the chip packing density for microprocessors [maximum number of transistors possible on an integrated circuit (IC)] would double every year. Mathematically stated,

$$\text{Chip packing density} = 2^{(\text{present year} - 1964)}.$$

As density of the circuits on a chip increases, more functions can be put on the chip. This reduces the communication path between logic elements and increases the system clocking. Thus, the processing power of individual ICs is increased.

On similar lines, in 1984 Bill Joy, one of the founders of Sun Microsystems, predicted that the processing speed of microprocessors, measured in million instructions per second (MIPS), would double every year. Mathematically stated,

$$\text{MIPS} = 2^{(\text{present year} - 1984)}.$$

Both these predictions seem to hold true to the present. The maximum capacity on a single chip has increased from less than 100 bits in the early 1960s to more than 200 Mbits at present. Similarly, the processing power of microprocessors has increased from less than one-tenth of MIPS to 800 MIPS at present. This has been accompanied by a remarkable reduction in cost for the same computing power.

To take an example from our daily lives, a standard PC with a 100-MHz Intel processor and 8 MB of RAM was available for $1,800 in 1995 in the United States. In 1999, for the same price one could get a PC with a 500-MHz Intel processor and 128 MB RAM. Moreover, one could get a PC with a 400-MHz Intel processor and 96 MB RAM for less than $1,000. PCs with comparable processors from companies other than Intel could be bought for even less.

These developments have put enormous computing power at very affordable prices in the hands of companies. They do not have to invest millions on buying (and then maintaining) mainframe computers. They can invest in much more affordable workstations that have enough computing power for their needs.

6.1.1.2. Networking Technology

By the mid-1980s, the main interfaces between the mainframe and the programmers were the "dumb" or "text-only" terminals. These were simple input/output devices that could talk only to the mainframe computer and not to each other. With the invention of PCs, there was a need for communication between computers. The earliest such computer networks included several computers using the same printers or storage devices. The next logical extension of this phenomenon was for computers to talk to each other in a computer network.

Computer networks consist of computers, cables (twisted cables, coaxial cables, fiber optic cables, etc.), and special data switches that control and route the information between different computers based on some established protocols. The following are the most popular forms of computer networks today:

- *Local area networks (LANs):* LANs are used for data exchange between computers in a small geographical area. LANs have dedicated communication channels and are typically owned, maintained, and operated by the end users. LANs have high transmission capabilities ranging up to 100 Mbits/sec.
- *Wide area networks (WANs):* WANs handle computer networks over long distances ranging from different office buildings to offices across a city or across the world. In such networks, the end user is responsible for telecommunication contents, protocols, and management, whereas the physical transmission devices (transmission cables/satellites, etc.) are provided by outside companies. Typical transfer rates are much slower than those of LANs, in the range of 64 Kbits/sec.

These networking capabilities enabled the evolution of new paradigms of computing — such as client/server technology and Internet technology — which are discussed in detail in Section 6.1.2.2.

Hardware development includes computer hardware development and networking hardware development. Computer hardware development can be highlighted by the following statistics:

> - The maximum capacity on a single chip has increased from less than 100 bits in the early 1960s to more than 200 Mbits at present.
> - The processing power of microprocessors has increased from less than one-tenth of MIPS to 800 MIPS at present. This has been accompanied by a remarkable reduction in cost for the same computing power.
>
> Furthermore, the developments in computer networks have been phenomenal. The most popular forms of computer networks are the local area networks for small geographical area and the wide area networks that can connect computers throughout the world.

6.1.2. SOFTWARE DEVELOPMENT

The evolution of computer software has closely followed the development of computer hardware. This development is discussed in two sections on system software and application software.

6.1.2.1. System Software

System software is a set of generalized programs that manage the resources of the computer, such as the central processing unit, communication links, and peripheral devices (Loudon and Loudon, 1995). System software has three main components: the operating system, language translators, and utility programs.

The operating system forms a software layer between the hardware and machine language. The operating system allocates and assigns system resources, schedules the use of computer resources and computer jobs, and monitors computer activities. It thus makes users and applications independent of the details of the hardware.

The evolution of operating systems followed the evolution of hardware. In the 1960s and 1970s, the mainframe operating systems were totally hardware dependent. They allowed a multiuser environment. However, they could not interface with systems running on other computers. Examples of these systems are MVS from IBM and VMS from Digital.

The growth in the number of computer manufacturers, programming languages, and software vendors in the 1970s fueled the need for an operating system that was independent of the hardware system. The UNIX operating system was an answer to these needs. It was independent of hardware and could be quickly adapted to different hardware environments. UNIX allowed users to interface to other systems (i.e., it had open interfaces). However, UNIX began to be influenced by hardware manufacturers and different versions of UNIX (which were only partially compatible) were developed. Examples include AIX (IBM), HP-UX (HP), and Solaris (Sun).

The arrival of desktops and personal computers resulted in the development of operating systems that fulfilled the need of these new breeds of computers. As opposed to the operating systems for mainframes and minicomputers, which were multiuser environments, the operating systems for desktops were single-user systems optimized to run such applications as word processors and spreadsheets. Examples of the desktop operating system include OS/2 from IBM and MS-DOS from Microsoft.

In all these systems, users interacted with the computers by typing commands. However, the development of the Star system (Johnson *et al.*, 1989) at Xerox laid the foundation for graphical user interfaces. The first such system to become available was the Macintosh from Apple computers in 1984. It was followed by the introduction of Windows for MS-DOS from Microsoft and Presentation Manager for OS/2 from IBM. Graphical user interfaces for UNIX were developed on the basis of X Windows (Jones, 1991) developed at MIT. The late 1990s saw the emergence of Microsoft Windows 95 as the most popular desktop operating system, which captured more than 80% of market share and was loaded onto hardware made by different PC manufacturers, including IBM, Compaq, Dell, Gateway, and Toshiba. An example of a graphical user interface based on an open operating system for a multiuser environment (for servers) is Windows NT by Microsoft.

The other two components of the system software — language translators and utility programs — have kept pace with developments in operating systems. Language translators such as compilers interpret the software written in high-level programming languages for the operating system. This makes programming languages independent of the hardware. The language translators of the 1990s have gone a step further. Compilers for object-oriented languages such as JAVA (developed by Sun Microsystems) tend to be independent not only of the hardware but also of the different types of operating systems. They are like virtual machines that enable the software to run on multiple platforms. Utility programs such as installers and uninstallers enable end users to install and remove applications from their computers without having deep knowledge of computer systems. This has made the use of computers more user friendly.

These developments in system software have paved the way for radical advances in application software. As a result, application software has been made more user friendly.

6.1.2.2. Application Software

Application software comprises computer programs that fulfill a business need and require little or no coding on the part of the end user. Application software ranges from powerful data processing systems used by the financial industry and scientists to computer-aided drawing (CAD) packages used in engineering and simple word processing software used in most offices and homes.

The evolution in application software has followed the evolution of hardware and system software. Application software in the 1970s was designed keeping in mind the optimal use of the expensive mainframe computers. The main input devices were dumb or text-only screens. The format of the display was determined by the programmers/developers and the end users had to be trained to use the format. The appearance of two different applications within the same department of a company could have an entirely different feel and need entirely different training. These computer applications were processed in a batch mode. The user would input the data for processing and submit it to the computer. The computer would then put this job in a queue along with requests from other users in a way so as to optimize the use of the computer. The users would have to wait for minutes or even hours before they got the results of their requests. Also, there was no way that the users could customize the input format to their liking.

The arrival of desktop computers in the early 1980s challenged this monolithic software architecture which depended so heavily on the mainframe computers. This was the beginning of the era of distributed computing. A new breed of software architecture emerged, which was called client–server architecture. The client–server architecture used the power of both the desktop computer and the more powerful mainframe computers. The application software was split into two main domains, the presentation domain and the computing domain. The presentation domain was responsible for displaying the input/output screens, the associated computing, and the interaction with other peripheral devices such as printers, and it was run on the less powerful machines. The computing domain was the part that did the heavy-duty computations and resided on the more powerful computers. The presentation domain would capture the computing requirements from the end users and was called the client. This client then placed the processing requirement on the computing domain, which was called the server. The processing power requirements were split between the two domains, thus freeing up computing time on expensive and overloaded mainframe computers.

Client–server architecture caused two very important changes in how computers were used in the industry. The first change was in the design of the user interfaces. Since the programs running the user interfaces did not reside on the mainframe and did not use their computing time, there was much more flexibility in how they could be designed. Thus, ease of use by the end users became one of the parameters in their design. End users began to have options for customizing the user interface. Users could choose from a set of options to make the appearance of the interface suit their style without affecting the functionality. This reduced the training time of users and increased their productivity. The second change was the response time to the requests. Since mainframe computing usage was freed up, they could process faster. Also, computing could be split between different smaller computers so that the queue for each of them was smaller. The total cost of these smaller computers was often less than the cost of a single large mainframe.

The coming of the desktops brought along with them an entirely new set of application software for desktop computing. The most important were word processing and spreadsheet software. This software resided on the desktops and did not require the use of mainframes at all. The number of software vendors exploded with the development of desktops, as did the number of computer users. The computer was no longer confined to large companies or research institutions. Smaller businesses could now afford these computers. The new breed of software vendors provided new users with application software for their business needs (that were not very computing intensive) that would run on the desktops.

The next major development in application software came with the introduction of the graphical user interfaces. The interaction with computers was made much more easier and intuitive by the graphical user interfaces.

The 1990s saw the emergence of two new trends in application software: object-oriented programming and the Internet. Object-oriented programming affected the application software developers more than the end users. Object-oriented technology uses software modules called objects to build the software. These objects are like pieces of Lego. Just as the same building blocks can be put together in different ways to create different things with Lego, object-oriented programming uses the same building blocks, or objects, to build different types of applications. This dramatically reduced the development time of applications and increased the richness of their functionality as programmers could reuse existing objects and did not have to "reinvent the wheel" every time.

The most recent development in applications was the Internet. The Internet enabled organizations to make information about their company available to people outside their company without giving them access to the company's proprietary systems. The natural extension of the Internet was an intranet that served as an electronic bulletin board on which different departments could post information in a central place that was accessible to all employees. The Internet has come a long way since the initial days. Current applications include buying and selling, running applications, video conferencing, and conducting training classes over the Internet.

> Software development includes system software development and application software development.
>
> UNIX was the most interesting attempt to provide an operating system independent from the hardware, and thus to allow users to communicate easily. The development of the Star system at Xerox laid the foundation of graphical user interfaces. Graphical user interfaces for UNIX were developed on the basis of X Windows developed at MIT. The late 1990s saw the emergence of Microsoft Windows 95 as the most popular desktop operating system.
>
> The last two major trends in application software were object-oriented programming (JAVA) and the Internet.

6.1.3. BUSINESS NEEDS

The 1980s and the 1990s saw the emergence of customers as the driving force for industries. Pressure on companies was increasing from all corners. Customers were demanding more customized products with high quality and low cost. It was becoming increasingly difficult for companies to cope with this situation using the established channels and rules of production and inventory management. Companies had to introduce new products in shorter intervals, and the products had shorter life cycles. Thus, on the one hand, product development time had to be decreased to meet these changes; on the other hand, the traditional concept of make to stock was becoming increasingly more inefficient. It was becoming more difficult to predict demand due to changing customer needs. Therefore, companies could not meet the fluctuations in demand by building inventory. There was the danger that sudden changes in customer needs would render the inventory obsolete. This new situation required a quick response to customer needs, without carrying excess inventory. Thus, communication of information from customers to retailers, manufacturers, and suppliers had to be extremely fast in order to respond to the customer in a timely manner. Computers provided an efficient way of attaining faster communication.

The explosion in the number of products manufactured by a company resulted in problems for the operations group. Production management became much more complex because of the vast number of products that were in process at any given time in the system. There were issues relating to planning the material and resources so as to optimize the throughput and minimize the work in process and the lead time (the two objectives oppose each other). Tracking the progress of different operations and materials through the system was becoming increasingly difficult. Thus, there was a need for sophisticated planning, control, tracking, and execution systems.

The increasing amount of data that companies were generating (and needed) had to be archived in a way that enabled easy retrieval. This was the driving force for enterprise-wide databases.

Companies with offices throughout a country wanted to enable the different offices to communicate with each other, use information/applications located in different geographical locations, and be linked with the central systems. Also, companies were becoming increasingly global, with operations in different parts of the world. This posed additional problems because each country has its own language and currency. Also, rules and regulations and work cultures differ between countries. Thus, an application software designed for use in one country could not be used throughout the company. Companies had to customize their software for the different countries while trying to keep them integrated. These requirements increased the need for client–server technology, which is becoming ever more popular.

The Internet is drastically altering the traditional channels of material movement. The traditional concept of vendors supplying raw materials to manufacturers, which would then sell manufactured items to wholesalers, which would then sell them to retailers and the retailers would sell them to customers is changing. At first, most companies viewed the Internet as a medium for making information available to their customers. However, it soon became a medium of buying and selling things, so much so that it threatened the very existence of brick-and-mortar retail stores in certain industries. A good example of a dominant retailer on the Internet is Amazon.com, which started selling books online and now sells music and movies among other things. People are beginning to question why manufacturers cannot remove the wholesalers and retailers from the supply chain and sell things directly on the Internet.

The Internet is emerging as one of the major areas for which companies need application software to participate in the next major paradigm shift in business. This is driving the growth of software development in this field.

> The needs of business are easy communication (Internet and intranet), the ability to manage huge databases, and application software that allows a high level of reactivity.

6.1.4. HUMAN RESOURCE DEVELOPMENT

One of the major requirements for any technology to become successful in the industry is the availability of human resources that are capable of using the technology. Use of computers in the 1960s and 1970s was restricted to programmers or highly trained people, both in industry and in academia. In academic institutions, the use of computers and teaching of computer skills were restricted mainly to the computer science departments. Computers slowly proliferated into other engineering and science departments that needed heavy computing power.

However, it was the desktops that revolutionized the use of computers. These machines were not very expensive so small businesses and even individuals could afford them. Thus, application software developers could now cater to an entirely different market. This market was much larger than the market dominated by mainframe applications. The developers created exciting new applications for desktops, including word processors, spreadsheets, presentation aids, simpler versions of programs that ran on mainframes (e.g., small databases), and even computer games. Thus, the computer found a place in offices for purposes other than heavy-duty computing. Even in academics, departments such as business and humanities began to use and teach the use of application software in a way very different from that taught in the engineering and computer science departments. The most important change that the desktops brought about was the fact that one did not have to be a programmer to use a computer. The use of computers empowered the common man in the field of computing just like owning public company stocks empowered the common man financially.

The next major advance in the use of computers was enabled by graphical user interfaces. These interfaces were more intuitive and easy to use than the interfaces for which one had to memorize and type commands. The ever-decreasing prices of computers were making it possible for increasingly more people to own computers.

The latest explosion in the use of computers was due to the Internet. The Internet has generated tremendous interest by many people, including those who do not have a technical background. The Internet has broken all boundaries. People can communicate across continents over the Internet (without having to pay for expensive international phone calls). They can exchange and access information on almost any subject under the proverbial sun and even beyond.

This information revolution has provided companies with a vast pool of human resources with different levels of expertise. Companies can now use various application software in their businesses without having to extensively train their employees. This availability of trained human resources is becoming one of the major enablers for businesses to adopt computer technology.

> Two factors increased the proliferation of computers: the emergence of the desktop computer and the Internet, which opened the world to everybody, including those without a technical background.

6.2. APPLICATION ENABLERS: DEVELOPMENTS AND STATE OF THE ART IN BUSINESS APPLICATIONS IN SUPPLY CHAIN MANAGEMENT

Business applications have advanced greatly from the COBOL-based packages for accounting and finance departments and material resource planning (MRP) packages for the manufacturing department to the Internet-based e-business solutions of today. The progress in the field of business application development has followed the following four stages:
- Legacy applications
- Enterprise resource planning (ERP) applications
- Supply chain planning (SCP) applications
- Internet business applications

6.2.1. LEGACY APPLICATIONS

Most of the business applications in the 1960s and 1970s were mainframe based and shared the characteristics of mainframe applications. Each functional department, such as finance, accounting, manufacturing, and marketing, had its own applications. Data was rarely shared between the different computer systems (except in cases in which the hardware was the same) and integration of applications between the different computer systems was as common as the unicorn. Cross-training of resources between the different applications was not in vogue. This trend continued in the 1980s, although the applications became increasingly sophisticated.

The finance and accounting departments had applications for maintaining finances, including expenses, accounts receivable/payable, and payroll. The marketing department used software applications to analyze market scenarios and for forecasting. The sales organization used software for entering and maintaining customer information as well as customer orders.

The engineering department used a class of sophisticated software called CAD software. These software applications allowed engineers and designers to design product components using two- and three-dimensional rendering. The resulting design data were also stored in databases maintained by the engineering organization.

The logistics organization used software applications to handle both inbound and outbound operations. For inbound logistics, software was used for procurement and inventory management. The software maintained databases for approved vendors, the on-hand inventory, and the material requests from the manufacturing organization. For outbound logistics, software was used for transportation planning and warehouse management.

The manufacturing organization used software for both planning and control. However, even though these applications were used by the same organization, they were not integrated and planning was often sequential instead of concurrent. Regarding planning, applications for production planning, master production scheduling (MPS), and rough-cut capacity planning were used at the tactical level. Materials requirement planning (MRP) and capacity requirement planning (CRP) were used at the operational level. Scheduling software was used at the execution level.

In the 1970s, these applications were stand-alone and planning was sequential. The customer order data and forecast were used to create the production plan, the output of which was fed to a master production scheduling application. Based on the output of MPS, an approximate capacity plan was formulated. The output of the MPS was fed to the MRP application, which created the detailed material plan. Based on this plan the CRP created the detailed capacity plan. If the material and capacity plans were incompatible, numerous iterations between the two had to be performed to obtain a compatible set of plans, which was often not attained. The detail schedule at the execution level on the shop floor was created based on the output of MRP and CRP. Tracking of parts on the shop floor was accomplished by the use of bar code scanners. The same technology was also used by the logistics group for tracking the inbound and outbound material.

The late 1970s and 1980s saw the maturation of these planning technologies into a more comprehensive set of applications called manufacturing requirements planning (MRP II) integrated these stand-alone applications under one umbrella. However, MRP II still maintained the sequential nature of planning, although the data flow was more integrated. Also, it maintained the dichotomy between material and capacity planning, although integration made the iterations easier and faster.

The different departments guarded their applications with zeal and data sharing between applications in the different departments was severely restricted by bureaucratic red tape as well as the inability of the applications to read data from different systems. The sharing of information between computer systems of different companies was unheard of.

The standardization in computer hardware and software interfaces resulted in the sharing of information between the different constituents of the supply chain. This was accomplished through electronic data interchange (EDI). In EDI, different companies agreed on a standard data format to transfer data between them. This format was often proprietary and access was restricted to partner companies. Once the format was established, it was very difficult to change. Also, partner companies had to invest a great deal in making their applications compatible with the agreed on format. Companies used EDI for placing procurement orders and other information such as inventory and sales data. Although it was a good beginning, EDI transactions were batch transactions and not done in real time. Also, the type of data shared was very limited. This was a far cry from the concepts of supply chain planning as described in this book.

6.2.2. ENTERPRISE RESOURCE PLANNING APPLICATIONS

A friend of ours sent us the following joke, which is supposed to portray the channel of communication in a company:

From: Managing Director
To: Executive Director
Tomorrow morning there will be a total eclipse of the sun at nine o'clock. This is something which we cannot see every day. So let the workforce line up outside in their best clothes to watch it. To mark the occasion of this rare occurrence, I will personally explain the phenomenon to them. If it is raining we will not be

able to see it very well and in that case the workforce should assemble in the canteen.

> From: Executive Director
> To: Departmental Heads
> By order of the Managing Director, there will be a total eclipse of the sun at nine o'clock tomorrow morning. If it is raining we will not be able to see it in our best clothes, on the site. In this case the disappearance of the sun will be followed through in the canteen. This is something we cannot see happening every day.

> From: Departmental Heads
> To: Sectional Heads
> By order of the Managing Director, we shall follow the disappearance of the sun in our best clothes, in the canteen at nine o'clock tomorrow morning. The Managing Director will tell us whether it is going to rain. This is something which we cannot see happen every day.

> From: Section Heads
> To: Foreman
> If it is raining in the canteen tomorrow morning, which is something that we cannot see happen everyday, the Managing Director in his best clothes, will disappear at nine o'clock.

> From: Foreman
> To: All Operators
> Tomorrow morning at nine o'clock, the Managing Director will disappear. It's a pity that we can't see this happen every day.

Exaggerated and funny though this joke may sound, it highlights a fundamental problem in the industry: that of data consistency as it passes through different channels. A company with modern computer systems could have avoided this comic situation by a direct e-mail from the managing director to all employees. This highlights other problems in the scenario of the joke, specifically those of bureaucracy, duplication of effort by middle management, and the time lag between the initial decision and the time when the intended receivers are informed of it.

These problems were shared by industries that had segregated the software applications of the different departments. To reiterate, the problems faced by companies having stand-alone applications were

- Duplication of data and effort: For example, the product data were maintained by both production and engineering departments.
- The availability of data to different departments was not in real time. For example, engineering changes to product design were not communicated to production in a timely manner, resulting in production of components that were obsolete.

6.2. APPLICATION ENABLERS

- Often there was distortion in transferring of data from one application/department to another. Frequently, the component names/codes were different in different departments. Thus, the same component could be referred to by three different names in the logistics, production, and engineering departments.
- The planning of the different departments was segregated. The departments did not have enough knowledge about the state of affairs of other departments. Therefore, planning by one department ignored constraints by other departments. The plans thus created were often incompatible and had to be revised repeatedly, wasting time and effort. For example, the sales organization promised order due dates based on historical data without taking into account the actual state of the shop floor. This caused orders to be manufactured and shipped late, resulting in unhappy customers and a tarnished image for the company.
- The applications did not support business process across internal departments and with customers/vendors.

The next evolution in the integration of information systems and applications in an organization was software applications called enterprise resource planning (ERP) systems. ERP systems claimed to be the provider of application software for the whole enterprise. They claimed to support business processes in different areas of an organization while maintaining integration between them. They also claimed to be a source of competitive advantage to companies.

As the name enterprise resource planning suggests, it evolved from MRP II and extended to include other departments of a company besides manufacturing, such as accounting and human resources. The ERP systems available today have evolved from different starting points. The market leader in ERP systems, SAP AG (Walldorf, Germany) started with MRP II and accounting solutions and then incorporated human resource management, logistics operations, and other industry-specific solutions. Its main customers have been large *Fortune* 1000 corporations. On the other hand, PeopleSoft (Pleasanton, CA) started with human resource management solutions and moved into accounting and manufacturing. Oracle (Redwood Shores, CA) was the leader in database systems before it moved into the ERP domain. Baan (Barneveld, The Netherlands) targeted midsize companies that could not afford expensive and lengthy implementations offered by other ERP vendors.

Although some of these systems have existed for 20 years, starting with mainframe-based applications, ERP systems became popular in the early 1990s. The main catalyst for this popularity was the introduction of client–server-based technology.

The most common architecture used in the ERP systems today uses a three-layer client–server architecture:

- Presentation layer for hosting the graphical user interface
- Application layer for implementing the application logic
- Database layer for storing and retrieving the business data

These layers are often run on separate computers and thus share the processing load among themselves, resulting in significantly improved response time. This also enables the use of relatively inexpensive computers because very heavy loads at each unit do not need to be supported. This architecture enabled the use of advanced functionality such as sophisticated graphical user interfaces and multimedia technology, interfaces to external applications, and high performance and scalability of the applications.

Buck-Emden and Galimow (1996) outlined the design principles behind the client–server-based R/3 ERP system by SAP as follows:

- Complete business processes with consistent quantity and value flow: The application covers business processes throughout the entire organization. The quantities and values of a process are represented consistently, up to the date in all business components at all times, and are available in real time.
- Data for a process is recorded only once: Once entered in the system, data is available to the different components of the system, thus avoiding duplication of effort and resources.
- Configurable business processes through integrated customizing functions.
- Support for business process reengineering.
- Independence of business solutions from system components.
- Internationalization of applications, where the presentation layers are customized for display in different languages and the application layer is designed to handle the rules and regulations in different countries, including currency conversions.

These principles, though specific to the SAP R/3 system, are representative of ERP systems in general. A comprehensive sample of the different application areas addressed in the ERP systems is shown in Table 6.1.

While supporting the different functional areas presented in Table 6.1, ERP applications maintain the information for these different areas in a central data warehouse. Thus, information from one functional areas is

TABLE 6.1 Application Areas in a Typical ERP System

Main application area	Subarea	Specific functions
Finance and accounting	General ledger accounting	Business accounting
		Balance sheet and profit planning
		Reporting and closing procedures
	Accounts receivable	Customer management
		Invoice and credit memos
	Accounts payable	Invoice receipt and posting of receipts
		Credits and down payments
	Funds management	Planning and budgeting
		Expense account management
		Fiscal accounting
	Financial controlling	Cash management
		Foreign exchange transactions
		Bank account clearing
	Financial asset management	Loans and stocks
		Borrowing
	Controlling	Cost center accounting
		Profit center accounting
		Profitability and market segment analysis
		Order cost accounting
		Product cost accounting
Sales	Orders processing	Order entry
		Order promising
		Shipping
		Invoicing
	Customer management	Managing customer information
		Quotations
		Contracts
Manufacturing	Planning	Production planning
		Master production scheduling
		Materials requirement planning
		Capacity planning
	Control	Production control
		Shop floor status data
		Quality control
	Maintenance	Maintenance planning
		Inspection planning
		Maintenance status tracking

(continues)

TABLE 6.1 (*continued*)

Main application area	Subarea	Specific functions
	Asset management	Depreciation simulations and computations
		Value assignments
		Asset history management
Logistics	Procurements	Material requirements
		Vendor information and evaluation
		Purchase orders
		Receiving
	Inventory management	Material tracking
		Material allocations
	Transportation	Transportation planning
		Shipment tracking
		Third-party logistics
	Warehouse management	Warehouse capabilities and structure
		Warehouse inventory
		Warehouse planning
Human resource management	Personnel	Personnel development
		Organization planning
		Workforce planning
		Time reporting
		Resource allocation
		Personnel information
	Accounting	Payroll accounting
		Expense tracking
		Employee benefits
Office applications	Communications	E-mails
		File transfers
		Address management
		Links to Internet and intranet
		Calendars
		Document sharing
	Personal computing	Word processing
		Spreadsheets
		Graphics and presentation materials

TABLE 6.2 Leading ERP Software Vendors

Vendor	Headquarters	Web site	1999 revenues (in millions of dollars)
SAP	Walldorf, Germany	www.sap.com	5110 (Euro)
Oracle	Redwood Shores, CA	www.oracle.com	9700
PeopleSoft	Pleasanton, CA	www.peoplesoft.com	1429
J. D. Edwards	Denver, CO	www.jdedwards.com	944
Baan	Barneveld, The Netherlands	www.baan.com	635
SSA	Chicago, IL	www.ssax.com	316
GEAC	Markham, Canada	www.geac.com	793
Intentia	Kista, Sweden	www.intentia.com	3062 (SEK)
QAD	Carpenteria, CA	www.quad.com	239
Lawson	Minneapolis, MN	www.lawson.com	270
Epicor Software	Irvine, CA	www.epicor.com	258
Interbiz	Islandia, NY	www.interbiz.com	Subdivision of Computer Associates
Wonderware	Newton, MA	www.wonderware.com	Subdivision of Invensys plc.
Mapics	Atlanta, GA	www.mapics.com	134
Infinium Software	Hyannis, MA	www.infinium.com	122
Ross	Atlanta, GA	www.rossinc.com	100
IFS	Linkoping, Sweden	www.ifsab.com	1948 (SEK)
Symix	Columbus, OH	www.symix.com	130

available to other areas as soon as it is entered in the data warehouse. The leading ERP software vendors throughout the world are listed in Table 6.2.

ERP systems created a revolution in the field of enterprisewide information systems with their capacity to integrate different functional areas of an enterprise. They also helped companies overcome their inertia and reengineer their business processes. They gained popularity with large organizations, and their popularity continues to increase. AMR Research predicted that the ERP market could be a $50 billion industry by 2002, growing at a rate of 37% from a market of $14 billion in 1998.

ERP systems seemed to offer a panacea for all company problems. However, they did not live up to that promise and also created new problems of their own. ERP application implementations were expensive and drawn out. Typical implementations took 18–48 months (Enslow, 1996), cost millions of dollars, and had a payback period of 2–5 years. These systems had hundreds of end users who had to be trained. Also, these new systems required an extensive support structure. This proved disruptive for companies as resources were taken away from day-to-day operations.

As the name enterprise resource planning suggests, these systems were designed primarily to integrate operations within an enterprise. Supporting interaction with other companies in the supply chain was not the primary design objective of these systems, which became apparent when companies tried to extend their ERP systems to supply chain planning. The ERP systems failed miserably.

The second main shortfall of the ERP systems is aptly captured by the following comment of a manager at a company with which we worked that implemented an ERP system: "What's missing in ERP is the P." Analysts at major application software market research firms agree (Enslow, 1996; Gormley *et al.*, 1997). Gormley *et al.* note four main limitations of ERP systems:

- *Transactions without responsiveness:* ERP systems did not support quick response to changes in demand, supply, labor availability, or machine capacity.
- *Production focus without understanding demand:* ERP systems treated demand forecasts as an external input. They failed to resolve the often divergent sales projections of marketing, operations, and sales. This resulted in excess inventory of some products and a shortage of others.
- *Control without intelligence:* ERP systems enabled companies to institute radical business process changes. However, some of these changes were based on old paradigms that were inefficient in the new scenarios. For example, production planning was still driven by MRP systems that assumed infinite resource capacity and planned for material and capacity separately.
- *Span without alignment:* Integrated ERP packages allowed a company to tie multiple plants and distribution facilities together from an organizational and inventory perspective. However, these systems could not leverage this information because they could not consider resources in these plants as alternates for each other and distribute the load accordingly.

These limitations presented an opportunity for a new class of application software known as supply chain planning software.

6.2.3. SUPPLY CHAIN PLANNING APPLICATIONS

ERP systems laid the foundation on which a new class of software applications emerged called supply chain planning (SCP) systems. These systems utilized the data-generating power of ERP systems to their advantage, and using this data provided planning functionality that was far superior to that offered by ERP systems. SCP systems were not dependent on ERP systems for data and could just as easily obtain the data from the legacy systems. However, the attention brought to the field of application software as a means of competitive advantage by ERP systems definitely helped emerging SCP applications. Soon, SCP applications emerged from the shadow of ERP systems from being just add-ons to being the primary application software for companies (Enslow, 1996; Gormley *et al.*, 1997). SCP systems are also called advanced planning systems (APS). This is a more appropriate name because these applications offer planning solutions only for the internal supply chain of a company. However, the acronym SCP has become popular and widely recognized.

SCP systems offered planning functionality that was far superior to that offered by ERP systems. They were differentiated from ERP systems on three counts:

- They were based on the latest technologies.
- They incorporated new planning paradigms.
- They were easier to implement, provided higher return on investment (ROI), required less resources, caused less disruption in day-to-day operations, and the payback period was smaller.

6.2.3.1. New Technologies

Most SCP software architectures are client–server based. They have graphical user interfaces that are designed with the various end users in mind. SCP software applications were developed using the latest tools available, namely, object-oriented programming, artificial intelligence, and genetic algorithms. These applications had open interfaces and since they relied heavily on legacy systems and/or ERP systems for data, they had interfaces that were easy to integrate with such systems. However, their main advantage over ERP systems was their response times. Most SCP applications were memory resident. They loaded all the data into the main memory of the computer. Thus, they did not have to repeatedly retrieve and input data into databases. Although individual retrievals and inputs from and to databases take a fraction of a second, when the applications have to do these operations millions of times, the difference is apparent. The memory resident attribute made SCP applications an order of magnitude faster than ERP systems. Thus, as opposed to MRP runs that used to take several hours (and therefore many companies used to run their MRP systems once a week during weekends), the planning engines of SCP applications could run in minutes. This enabled planners to run these planning tools more often. They could now compress their planning cycle from weekly to daily or even multiple times a day. This allowed them to be more responsive to the changes in the system.

6.2.3.2. New Planning Paradigms

ERP systems were able to integrate applications across the enterprise very well. However, the applications they integrated were based on paradigms of the 1970s and 1980s. The supply chain planning applications also brought new planning paradigms. For the first time, these applications presented tools that provided feasible solutions matching supply and demand while taking into account system constraints. These constraint-based applications analyzed trade-offs between resource and labor availability, on the one hand, and cost of materials, labor, capacity, and transportation, on the other hand, to meet customer demand in a timely and cost-effective manner.

6.2.3.3. Implementation

As opposed to an implementation time of 18–48 months for ERP systems, SCP systems have an average implementation time of 9 months (Enslow, 1996). They are also less costly. The return on investment on these systems is high and the payback period is small.

6.2.3.4. Supply Chain Planning Solutions

SCP software emerged in the early 1990s. In the beginning, there was no one vendor or product that covered the whole breadth of the supply chain. The supply chain solutions could be divided into the following domains:

- Demand planning
- Supply planning
- Logistics planning

Different software vendors offered solutions in different domains. However, today individual software vendors have moved into more than one domain through in-house product development and/or acquisitions. The leading SCP vendors, such as i2 Technologies (Fig. 6.1) and Manugistics,

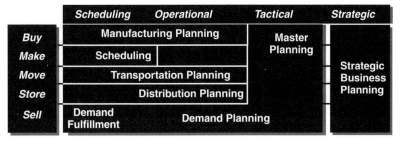

FIGURE 6.1 Different solution areas in supply chain planning (reproduced with permission from i2 Technologies).

now claim to offer end-to-end solutions for supply chain management. Next, we discuss the solutions in these three domains in greater detail.

Demand Planning

The sophisticated modeling and statistical techniques offered by these applications helped planners improve their demand projections. They could synthesize demand projections from different areas of the organization, such as marketing, sales, and manufacturing, to create a single uniform view of the demand. The areas addressed by these applications include demand creation, demand forecasting, and forecast netting. Users can have different views of the demand (e.g., by product, by customer, or by region).

Supply Planning

Supply planning is one of the strongest areas for SCP applications. They provide solutions at the tactical, operational, and execution level. Tactical solutions help the company answer questions such as

- Which factory should make what product?
- From which distribution centers should one source a demand?
- Which customers should have priority?
- How should one allocate scarce material and capacity?
- Where should one source purchase parts?
- What alternate parts could one use?
- What transportation mode should be used?
- What are alternative strategies for the previous questions?

At the operational level, the planning solutions are capable of generating feasible plans for material and capacity simultaneously. This is a major improvement from MRP systems that generated the material and capacity plans separately and then had to perform several iterations to make the two plans compatible. In generating such plans, these systems take into account the capacity constraints in terms of the processing rate and available time on the resource, resource calendars, and any anticipated maintenance or down times. They also take into account the availability of materials (from work-in-process inventory, raw material inventory, or components from vendors). They offer advanced material allocation and synchronization of components so as to satisfy customer demand in an optimal way. For example, in traditional MRP, if a product needs three components and one is delayed due to problems with a vendor, then the other two components will have to remain in inventory until the third part becomes available. However, SCP applications provide the capability to reallocate these two parts to meet some other customer demand. These systems also offer advanced features such as available-to-promise and real-time due date quoting. Thus, salespeople can promise delivery dates based on the dates provided by these systems that take into account the actual state of the supply chain. These dates are more realistic than the "standard" lead times most companies use in such situations.

At the execution level, the scheduling tools of these applications provide optimal schedules based on the plans provided by the tools at the operational level. These are at the level of minutes/seconds and can be modified very quickly based on the actual data on the shop floor.

Logistics Planning

The logistics solutions provided by SCP software vendors include tools for transportation planning, distribution planning, and warehouse management. Transportation planning tools enable companies to plan the optimum way to transport products from one location to another. These systems take into account the global capabilities of the transportation system and the requirements imposed by different locations. This global view enables these applications to consolidate shipments if possible and perform trade-off analyses between the costs of different modes of transportation. The distribution management and warehouse management tools help planners decide the best way to maintain warehouses and distribution centers in order to meet customer demand in a fast and cost-effective way. Some of the leading vendors of supply chain planning solutions are presented in Table 6.3.

The SCP applications discussed previously have taken planning capabilities of companies to a new level. They are becoming increasingly popular among companies due to their high ROI and short payback periods. Forrester Research predicts that the supply chain planning software industry will be a $18.6 billion industry by 2003, close on the heels of the ERP

6.2. APPLICATION ENABLERS 149

TABLE 6.3 Leading Supply Chain Planning Software Vendors

Vendor	Headquarters	Web site	1999 revenues (in millions of dollars)
i2 Technologies	Dallas, TX	www.i2.com	571
Manugistics	Rockville, MD	www.manugistics.com	152
Numetrix	Toronto, Canada	www.numetrix.com	Subdivision of J. D. Edwards
Logility	Atlanta, GA	www.logility.com	33
Synquest	Atlanta, GA	www.synquest.com	Privately held
Thru-put	San Jose, CA	www.thru-put.com	Subdivision of Mapics
Aspen Technologies	Cambridge, MA	www.aspentech.com	220
Adexa	Los Angeles, CA	www.adexa.com	Privately held
SCT/Fygir	Malvern, PA	www.sctcorp.com	466
Webplan	Ottawa, Canada	www.webplan.com	Privately held
Mercia	London, England	www.mercia.com	Privately held
SAP	Walldorf, Germany	www.sap.com	5110 (Euro)
Oracle	Redwood Shores, CA	www.oracle.com	9700
PeopleSoft	Pleasanton, CA	www.peoplesoftt.com	1429
J. D. Edwards	Denver, CO	www.jdedwards.com	944
Baan	Barneveld, The Netherlands	www.baan.com	635

industry. As noted earlier, a more appropriate name for these applications is APS. Even though some vendors claim to offer end-to-end solutions, the way these solutions have evolved has caused them to be an assembly of different products rather than a single solution. This modularity has both positive and negative aspects. Although modules are much easier to implement as stand-alone applications that are fully functional, they are more difficult to integrate into a single application implementation. Furthermore, SCP applications aim to optimize the internal supply chain of a company. They have very limited capability for enabling planning across multiple enterprises. Although some vendors have developed solutions for collaborative planning between companies, these are far from mature.

The newest applications emerging in business computing are Internet business applications, which enable a company to interact more closely with external sources and relate to concepts of supply chain as presented in this book.

6.2.4. INTERNET BUSINESS APPLICATIONS

The Internet has its origins in the U.S. Department of Defense initiatives of linking its multiple organizations. However, the Internet as we know it today, the World Wide Web, or the more familiar "www," emerged from Geneva. From its early days as the playground for computer jockeys to the present craze of dot.com software companies, it has evolved considerably.

E-commerce and e-business are the latest buzz words that have forced even traditional companies, such as Barnes and Noble (book retailers), Merrill Lynch (financial services), Sears (retailers), and IBM (computer hardware and services) in the United States; Nokia (phones) and Tesco (supermarket) in Europe; and Sony (electronics) and Toshiba (electronics) in Asia, to adopt these new paradigms as a way of doing business.

The first and most popular use of Internet applications in business was providing services (such as stock trading and financial advice) and goods (books, music, toys, etc.) directly to customers. The Internet provided a powerful tool and removed a barrier to entry for newcomers in retailing. Small businesses did not have to invest in real estate in prime locations and they were immediately visible and accessible to people throughout the world. However, this euphoria was short lived because the number of vendors on the Internet grew exponentially. It became increasingly more difficult for customers to find what they were looking for. Similarly, it became challenging for vendors to attract and retain customers. Since most of the products sold on the Internet were standardized products, cost was one of the main considerations. However, soon the online prices of goods stabilized and cost was no longer a distinguishing feature.

In this chaos, customer management emerged as one of the main tools for attracting and retaining customers. Online vendors had to adapt the retailing principles applied in brick-and-mortar stores to the Internet. Thus, they had to find the counterparts of attractive lighting, strategically located displays, energetic salespeople, etc., on the web. To fulfill these needs, software vendors emerged that provided software solutions for Internet-based customer relationship management (CRM).

CRM vendors provided retailing companies with applications that had user-friendly interfaces. These interfaces were simple and helped users find what they were looking for. These applications provided a customized shopping experience for users. Users could customize the interface to their liking, and this interface would be restored when the customer returned. These applications kept track of shopping habits of customers and guided them toward products that they liked or had bought in the past. For example, if a customer bought a book on supply chain management, the software

TABLE 6.4 Leading CRM Application Software Vendors

Vendor	Headquarters	Web site	1999 revenues (in millions of dollars)
Siebel	San Mateo, CA	*www.siebel.com*	790
Vantive	Santa Clara, CA	*www.vantive.com*	Subdivision of PeopleSoft
Clarify	San Jose, CA	*www.clarify.com*	Subdivision of Nortel Networks
Trilogy	Austin, TX	*www.trilogy.com*	Privately held
Broadvision	Redwood City, CA	*www.broadvision.com*	115

would provide him or her with other titles on supply chain management in the store at that time, provide a list of books that other shoppers bought who also bought this particular book, or provide the names of other books by the same author. The leading vendors of CRM application software are shown in Table 6.4.

The newest Internet-based applications are business-to-business applications, popularly called e-business applications. The customers in this case are not individual buyers but other businesses. This distinction between e-commerce and e-business is based less on the actual difference between the technologies used in the two scenarios and more on how the terminology emerged. One of the major differences in the case in which the transaction is business-to-business is in the procedures involved in buying. Typically, an individual customer goes on the web, searches for the vendor offering the best deal, buys the product, and pays using his or her credit card. However, when businesses are involved there are more technicalities involved. Often, businesses invite quotations, someone from the customers' side needs to verify that products being bought are of acceptable standards, invoices are created for payments, delivery modes are determined, etc. Thus, business-to-business transactions are more challenging not only technically but also because of the accompanying business processes.

However, e-business is currently the hottest new paradigm in business. Many companies are moving to adopt this paradigm. In addition to the new players that are developing solutions to provide the application software needed for enabling this paradigm, established players in ERP, SCP, and CRM are also jumping in, alone or in partnerships with other application vendors.

> Progress in business applications had followed the following stages:
> - Legacy applications that were enabled by the standardization in computer hardware and software interfaces: Communication between these applications was accomplished mainly through electronic data interchange.
> - Enterprise resource planning (ERP) applications that aimed to (i) reduce duplication of data and effort, (ii) make data available to different departments in real time, and (iii) remove distortion in transferring of data from one application/department to another.
> - Supply chain planning applications that offered (i) new planning paradigms, (ii) better return of investment, and (iii) greater efficiency due to the use of new technologies.
> - Internet business applications, such as e-commerce, business-to-business applications, and customer management, that allow companies to do business over the Internet.

6.3. CONCLUSION

From a technical standpoint, supply chains take advantage of hardware development (increasing capacity of chips and incredible possibilities of networking technology) and the development of system and application software. They are also influenced by business needs, which require flexibility and reactivity, and the evolution of human resources, which also become increasingly flexible due to the available resources.

In addition to these improvements, many applications have been introduced, tested, and improved. They are now available to facilitate finance, accounting, sales, manufacturing, logistics, human resource management, and office activities. In particular, supply chain-oriented applications are now available, and the Internet is playing an increasingly important role in the domain.

REFERENCES

Bowersox, D. J., Closs, D. J., and Hall, C. T., "Beyond ERP—The storm before the calm," *Supply Chain Management Rev.* **1**(4), 28–37, 1998.

Buck-Emden, R., and Galimow, J., *SAP R/3 System: A Client Server Technology.* Addison-Wesley, Harlow, UK, 1996.

References

Enslow, B., "Which comes first: ERP or supply chain planning projects," Gartner Group Report TV-000-109, 1996.

Gormley, J. T., III, Woodring, S. D., and Lieu, K. C., "Supply chain beyond ERP," in Packaged Application strategies, The Forrester Report, 1997.

Johnson, J., Roberts, T. L., Verplank, W., Smith, D. C., Irby, C., Beard, M., and Mackey, K., "The Xerox Star: A retrospective," *IEEE Computer* **22**, 9, 1989.

Jones, O., *Introduction to the X Window System.* Prentice Hall, Englewood Cliffs, NJ.

7

Conclusion

In an ideal supply chain, products and information flow freely. This leads to a dramatic reduction of the production cycle and, as a result, great flexibility and reactivity to cope with changing customer demands. A supply chain's performance can be further improved by introducing mechanisms that allow managers to anticipate emerging trends and customer needs, even before the customers become aware of them, in order to introduce products in the market at a time as close as possible to when customer demand originates.

The advantages of an efficient supply chain are well-known. The principle of handling new opportunities and orders as projects with cross-functional teams as opposed to being organized as functional departments is now widely accepted. Basic rules have been established to design production systems that are similar to the supply chain definition proposed in Chapter 2. As mentioned in Chapter 4, the goals of these rules are to

- Minimize the time required for converting orders into cash
- Minimize the total work-in-process
- Improve pipeline visibility, that is, make information available in real time to each of the participants of the supply chain
- Improve visibility of demand
- Improve quality
- Reduce cost
- Improve services

Nevertheless, none of the existing production systems fully fit the supply chain definition. A supply chain should be an organization in which partners closely cooperate in order to maximize the efficiency of the whole system. Cooperation is a notion that assumes equal responsibility and thus equal importance and power of each partner when a strategic decision has to be made. In the systems analyzed in Chapter 2, which are usually considered supply chains, we rarely observed a real cooperation but, rather, systems under the power of a dominant partner or even systems working on a master–slave basis. Such systems are more beneficial to the dominant partner than to the dominated ones. As a consequence, the growth of dominated partners is bounded, which leads to the impoverishment of the whole system in the long term. Indeed, some dominant partners are concerned about the efficiency of the other partners and help them to preserve or even increase their technical level by training employees, providing specific resources (machines, software, services, etc.), or developing new technologies. However, in times of economic recession, the dominant partner usually does not hesitate to modify the strategy of the system to preserve its position in the market, even if this strategy forces some partners out of business.

This situation, which weakens all partners including the dominant ones, occurs because of a lack of strategic thought on how a supply chain must be organized to maximize the efficiency of each of the partners. As a matter of fact, the so-called supply chains mentioned in the literature have been designed by companies that owned the information about the market and/or the technology. Starting from this dominant position, they first identified subcontractors to deal with increasing demands and, progressively, strengthened the links between the partners and increased the efficiency of the system by removing the barriers that slow the flow of information and products. Thus, moves toward supply chain architectures were constrained by the strategy of the dominant partners that took control of the system. Finally, the potential of the existing supply chains is restricted to the field of interest of the dominant partners. In particular, financial risks and benefits are not freely shared among the partners, thus affecting the innovative spirit of the whole system.

Thus, we claim that the design of a supply chain should start at the strategic level by defining a set of well-formulated rules to ensure that each partner will have the same interest and the same power in the management of the supply chain. The following rules should be included in this set:

- Strategic decisions should be made with the agreement of all partners. Reengineering, development of new products, switching

to a new generation of resources, fundamental changes in the marketing strategy, or selection of new suppliers or retailers are examples for which strategic decisions must be made.
- A sharing process, in which losses and benefits among the partners are fairly shared should be designed for each project. This sharing process is important not only for partners belonging to different companies but also for different divisions of the same company. In any case, it is helpful to evaluate the global return of investment at the level of each activity. It also encourages investments that improve the efficiency of the whole system, disregarding the local benefits or losses that result from a given investment.

It is important to focus on another aspect that penalizes most of the supply chains and that is rarely mentioned in the literature. Since production systems are progressively organized to be managed as a series of projects, research investments often appear to be too large to be included in the budget for the project. Furthermore, research concerns the long term, whereas a project is often planned in the medium or short term. Thus, research is either limited to technical adjustment or is outsourced. As a consequence, research competencies are rarely considered as important resources. Instead, only technical or short-term applied research is included in projects. Thus, innovation spirit gradually vanishes or is restricted to limited technical innovations, which leads to a drastic decrease of the competitiveness of the system. We observed this evolution in some steel and textile companies.

Since supply chains are targeted to generate returns on investment in the short term, long-term investments tend to be ignored. It is thus of utmost importance to enrich supply chains by introducing two types of components:

- A team that is in charge of tracking the evolution of research in the domain that directly concerns the supply chain as well as in related domains.
- A team that is in charge of transferring and adopting results of external research in the supply chain or developing comparable technologies within the supply chain.

> A supply chain is organized along projects. The goal is to reduce costs, improve services and quality, and reduce the time to convert orders into cash. The goal of a supply chain is to increase customer satisfaction and shareholder value.

Nevertheless, we think that none of the existing systems fit perfectly with the definition of an efficient supply chain. At least two additional aspect should be taken into account to further improve the supply chains:

- Think in terms of global efficiency when designing a supply chain. In particular, ensure that all partners have the same amount of power and responsibility, and introduce in each project a sharing mechanism that ensures that the risks and rewards are fairly shared among the partners.
- Pay particular attention to research since this activity affects the future of the supply chain.

APPENDIX

A.1. INTRODUCTION

During the past 5 years, pressure of the competitive market and new information technologies have affected the structures of production systems, calling for total quality, customer satisfaction, drastic reduction of time to market, manufacturing integration with R&D and marketing, and worldwide strategy and alliances. The supply chain paradigm is a result of this evolution. A supply chain is a global network of organizations that cooperate to improve the flow of material and information from suppliers to customers at the lowest cost and the highest speed.

A supply chain is organized by projects instead of by departments. Obviously, the goal of scheduling activities in such a system is no longer to optimally schedule a set of tasks in order, for instance, to minimize makespan or some criterion related to due dates. The objective is now to schedule tasks online, in the order in which demands appear in the system, and simultaneously to inform clients when their demands will be met. In other words, when an order arrives in the supply chain, the management system should be able to schedule all the tasks related to the order, from raw material to delivery, in a few seconds so as to minimize the completion time while keeping the work-in-process (WIP) at a reasonable level. In a supply chain, completing an order is a project. This project requires different types of resources, such as machines, transportation resources, workers, engineers, computers, and raw material. Scheduling the tasks that comprise the project consists of determining how best to use the idle periods of the resources while keeping the WIP as low as possible.

We first introduce the concept of a no-wait schedule, which is a schedule in which an operation starts as soon as the previous operation on the same part (if any) is completed. We first consider systems in which parts are manufactured by performing a sequence of operations (linear manufacturing systems), and we consider a generalized flow shop, which is a manufacturing system in which each part undergoes the same sequence of operations. Furthermore, an operation of a given type can be performed using identical resources, and two operations of the same type may require different manufacturing times when performed using the same resource. Limited flexibility exists since manufacturing times can be increased within certain limits at the expense of the unavailability of the resource required for the operation. These aspects are discussed in-depth in Section A.2. We then generalize the approach to systems that include assembly and disassembly operations (see Section A.3). Section A.4 demonstrates how the previous approaches can be used for real-time scheduling of production systems while controlling WIP.

Due to the interval-valued processing times, the shortest processing time (SPT) online rule used for traditional scheduling and sequencing (Baker, 1974) is not applicable here. Song *et al.* (1993) considered a similar problem of scheduling identical parts in a chemical processing tank line. They suggested an online heuristic called the earliest start time (EST) rule, according to which the start of each job entering the system is assigned to each machine as early as possible. This rule gives better results if combined with a combinatorial enumeration of variants in the case of conflicts. Generalizing this approach, we will study the case of the nonidentical part schedule and consider, instead of the classical SPT and EST rules, a new forward–backward earliest start time (FBEST) rule that has been proven to provide an exact optimum (the minimum makespan) for each job.

Callahan (1971) used a queuing model to study no-wait processing in the steel industry. Chu *et al.* (1998) and Chauvet *et al.* (1997) considered a surface treatment process, which is a no-wait problem. McCormick *et al.* (1989) studied a cyclic flow shop with buffers, which can be transformed into a blocking problem by considering buffers as resources with totally flexible processing times (i.e., processing times that can have any value between 0 and $+\infty$). Hall and Sriskandarajah (1996) provided a survey on scheduling problems with blocking and no-wait, and Rachamadugu and Stecke (1994) classified the flexible manufacturing system (FMS) scheduling procedures. This appendix should be considered as an intermediate approach: Some flexibility exists in the model in terms of the processing times, but this flexibility is limited.

For the sake of simplicity, we restrict these approaches to the manufacturing cycle and consider the machines as the only resources, but their extension to the process that covers customer demand to delivery is straightforward.

A.2. THE LINEAR MANUFACTURING SYSTEM

A.2.1. PROBLEM SETTING

In the problem, m types of operations are considered. For $i = 1, \ldots, m$, n_i is the number of identical resources (e.g., machines) that can perform an operation of type i. The problem arises at time 0. At this time, we know the idle periods (also called windows) for each resource.

For $i = 1, \ldots, m$ and $j = 1, \ldots, n_i$, $a_{i,j}^k$ is the beginning of the kth idle period of resource j able to perform an operation of type i. We assume that $k = 1, \ldots, f_{i,j}$. Similarly, $b_{i,j}^k$ is the end of the kth period.

We denote by θ_i^g and $\theta_i^g + \delta_i^g$, respectively, the lower bound and the upper bound of the time that a part that is manufactured following the manufacturing process g needed to perform an operation of type i. Indeed, $\theta_i^g > 0$. In the remainder of the appendix, we do not use the notation g for the sake of simplicity since we consider only one part at a time.

Furthermore, the m different types of operations mentioned previously are those visited by the part under consideration, in the order of their index, and the other resources are neglected. Indeed, the next part arriving in the system may visit another sequence of resources.

We denote by x_i the instant the part starts operation of type i. Since a part is not allowed to wait between two operations, x_i is also the completion time of the previous operation, if any. The x_i values are the solution of the problem. We assume that the part visits successively resources $1, 2, \ldots, m$. x_{m+1} is the completion time of the mth operation and thus the completion time of the product.

In this appendix, $F_{i,j}^k = (a_{i,j}^k, b_{i,j}^k)$ is the kth window concerning resource j dedicated to type of operation i. For resources dedicated to the same type of operation i, the windows $F_{i,j}^k$ are ordered in the increasing order of $a_{i,j}^k$ and, if two $a_{i,j}^k$ are identical, in the decreasing order of $b_{i,j}^k$. After these ordering processes, we have the ordered sets

$$S_i = \{(a_{i,j}^k, b_{i,j}^k)\}, \quad j = 1, \ldots, n \quad \text{and} \quad k = 1, 2, \ldots$$

We denote by $\alpha_{i,r}$ (and $\beta_{i,r}$), for $r = 1, \ldots, q_i$, the lower bound and the upper bound, respectively, of the idle periods of the resources performing operation i, ordered in the increasing order of the lower bounds and, in case of equality of the lower bounds, in the decreasing order of the upper bounds. According to the previous notation, a period $[\alpha_{i,r}, \beta_{i,r}]$ corresponds to one of the periods $[a_{i,j}^k, b_{i,j}^k]$, with these periods being ordered as mentioned previously. Thus,

$$S_i = \{[\alpha_{i,r}, \beta_{i,r}]\}, \quad r = 1, \ldots, q_i$$

where q_i is the number of idle windows associated with operation i. Note that in this kind of problem, β_{i,q_i} is always equal to infinity. Thus, a feasible solution always exists.

The problem can be set as follows: Find $x_1 < x_2 < \cdots < x_m < x_{m+1}$ and a sequence of intervals $SQ = \{S_i\}_{i=1,\ldots,m}$ which minimizes x_{m+1} and satisfies

$$\theta_i \leq x_{i+1} - x_i \leq \theta_i + \delta_i$$
$$x_i \geq \alpha_{i,r_i}$$
$$x_{i+1} \leq \beta_{i,r_i}, \quad \text{for } i = 1, \ldots, m$$

This set of constraints can be replaced by

$$x_{i+1} \geq \theta_i + x_i \tag{A.1}$$

The starting time of operation $i + 1$ should be greater than or equal to the minimal ending time of operation i.

$$x_i \geq \alpha_{i,r_i} \tag{A.2}$$

Operation i should not start sooner than the low bound of windows S_i.

$$x_{i+1} \leq \theta_i + \partial_i + x_i \tag{A.3}$$

The starting time of operation $i + 1$, which is also the ending time of operation i, should be less than or equal to the maximal ending time of operation i.

$$x_{i+1} \leq \beta_{i,r_i} \tag{A.4}$$

Operation $i + 1$ should not start later than the upper bound of windows S_i, for $i = 1 \ldots, m$. A sequence $x_1 < x_2 < \cdots < x_m < x_{m+1}$ is said to be feasible for the sequence SQ of windows if it verifies constraints (A.1)–(A.4).

A.2.2. AN FBEST ALGORITHM

A.2.2.1. Sequence Building

We assume that the time to move a part from one resource to the next is negligible.

Let us build the sequence $X = \{x_1, x_2, \ldots, x_m, x_{m+1}\}$ as follows, after choosing a sequence $[\alpha_{i,r_i}, \beta_{i,r_i}]$ of idle windows:

$$t_1 = \alpha_{1,r_1} \tag{A.5.1}$$
$$t_i = \max(\alpha_{i,r_i}, \theta_{i-1} + t_{i-1}), \quad \text{for } i = 2, 3, \ldots, m \tag{A.5.2}$$
$$t_{m+1} = \theta_m + t_m \tag{A.5.3}$$
$$x_{m+1} = t_{m+1} \tag{A.5.4}$$
$$x_i = \max(t_i, x_{i+1} - \theta_i - \delta_i), \quad \text{for } i = m, m-1, \ldots, 1 \tag{A.5.5}$$

In Result 1, we show that the elements of the sequence $x_1, x_2, \ldots, x_m, x_{m+1}$ verify inequalities (A.1)–(A.3).

A.2.2.2. Feasibility

Result 1

The sequence X defined by Eqs. (A.5.1)–(A.5.5) verifies inequalities (A.1)–(A.3) for the given sequence $[\alpha_{i,r_i}, \beta_{i,r_i}]$, $i = 1, \ldots, m$, of idle windows that have been selected.

Proof

a. X verifies inequality (A.1), i.e., $x_{i+1} \geq \theta_i + x_i$, for $i = 1, \ldots, m$.
   From inequality (A.5.5), we derive

$$\theta_i + x_i = \max(t_i + \theta_i, x_{i+1} - \delta_i) \leq \max(t_i + \theta_i, x_{i+1}), \quad \text{for } i = 1, \ldots, m \tag{A.6}$$

However, according to Eqs. (A.5.2) and (A.5.3),

$$t_i + \theta_i \leq t_{i+1}, \quad \text{for } i = 1, \ldots, m \tag{A.7}$$

Thus, using inequality (A.7) in Eq. (A.6):

$$\theta_i + x_i \leq \max(t_{i+1}, x_{i+1}), \quad \text{for } i = 1, \ldots, m \tag{A.8}$$

However [see Eqs. (A.5.4) and (A.5.5)],

$$t_i \leq x_i, \quad \text{for } i = 1, \ldots, m+1 \tag{A.9}$$

Then, using Eqs. (A.8) and (A.9), we obtain

$$\theta_i + x_i \leq x_{i+1}$$

and X verifies inequality (A.1) for the sequence of idle windows under consideration.

b. X verifies inequality (A.2), i.e., $x_i \geq \alpha_{i,r_i}$, for $i = 1, \ldots, m$. According to Eqs. (A.5.1) and (A.5.2),

$$t_i \geq \alpha_{i,r_i}, \quad \text{for } i = 1, \ldots, m \tag{A.10}$$

Considering Eqs. (A.9) and (A.10), we obtain

$$x_i \geq \alpha_{i,r_i}, \quad \text{for } i = 1, \ldots, m$$

and X verifies inequality (A.2) for the sequence of idle windows under consideration.

c. X verifies inequality (A.3), i.e., $x_{i+1} \leq \theta_i + \partial_i + x_i$, for $i = 1, \ldots, m$. According to inequality (A.5.5),

$$x_{i+1} - \theta_i - \delta_i \leq x_i, \quad \text{for } i = 1, \ldots, m$$

and thus

$$x_{i+1} \leq x_i + \theta_i + \delta_i, \quad \text{for } i = 1, \ldots, m$$

X verifies inequality (A.3) for the sequence of idle windows under consideration.

Result 1 does not mean that a sequence X defined by applying Eqs. (A.5.1)–(A.5.5) starting from a given sequence of idle windows is feasible. It shows only that X verifies three (of four) conditions to be feasible. We still have to verify condition (A.4).

The following definition is used in Result 2. A sequence $S^* = \{\alpha_{i,r_i}, \beta_{i,r_i}\}$ is a subsequent sequence of $S = \{\alpha_{i,v_i}, \beta_{i,v_i}\}$ if $v_i \geq r_i$ for $i = 1, 2 \ldots, m$ and $v_i \neq r_i$ for at least one $i \in \{1, 2, \ldots, m\}$.

Result 2

Assume that, in a given sequence X defined by applying Eqs. (A.5.1)–(A.5.5) starting from a given sequence S of idle windows, there exists $w \in \{1, \ldots, m\}$ such that $x_{w+1} > \beta_{w,r_w}$, where $[\alpha_{w,r_w}, \beta_{w,r_w}]$ is the wth window in the sequence of idle windows under consideration; then, no

A.2. THE LINEAR MANUFACTURING SYSTEM

subsequent sequence S^* of idle windows containing $[\alpha_{w,r_w}, \beta_{w,r_w}]$ will lead to a sequence X^* which is a feasible solution to our problem. Indeed, X is not feasible either since $x_{w+1} > \beta_{w,r_w}$ means that relation (A.4) does not hold.

Proof

Applying Eqs. (A.5.1)–(A.5.3) to S^* leads to values t_i^* which are greater than or equal to the vales t_i obtained by applying the same relations to S since $\alpha_{i,r_i}^* \geq \alpha_{i,r_i}$.

Considering Eqs. (A.5.4) and (A.5.5) and the fact that $t_i^* \geq t_i$ for $i = 1, \ldots, m$, we see that $x_i^* \geq x_i$ for $i = 1, \ldots, m$. Thus, $x_{w+1}^* \geq x_{w+1} > \beta_{w,r_w}$ and X^* is not feasible since Eq. (A.4) does not hold.

A.2.2.3. Real-Time Algorithm

In this section, we propose a real-time algorithm that leads to a feasible solution. We have proven that this feasible solution is optimal in the sense that it guarantees that the completion time of the product at hand is minimal, taking into account the available windows.

Algorithm 1

1. For each type of operation i, we order the idle windows as explained previously. The resulting sequence of idle windows is denoted by $[\alpha_{i,r}, \beta_{i,r}], r = 1, \ldots, q_i$, for $i = 1, \ldots, m$.
2. We set $r_i = 1$, for $i = 1, \ldots, m$. r_i is the rank of the window in which we try to perform operation i.
3. We build the sequence $X = \{x_1, x_2, \ldots, x_{m+1}\}$ by applying the following sequence of equalities: $t_1 = \alpha_{1,r_1}$, $t_i = \max(\alpha_{i,r_i}, \theta_{i-1} + t_{i-1})$, for $i = 2, 3, \ldots, m$; $t_{m+1} = \theta_m + t_m$, $x_{m+1} = t_{m+1}$, $x_i = \max(t_i, x_{i+1} - \theta_i - \delta_i)$, for $i = m, m-1, \ldots, 1$.
4. If, whatever $i \in \{1, \ldots, m\}$, inequality $x_{i+1} \leq \beta_{i,r_i}$ holds, we stop the algorithm.
5. Otherwise, for all $i \in \{1, \ldots, m\}$ that verify $x_{i+1} > \beta_{i,r_i}$, we set $r_i = r_i + 1$ and return to step 3.

Since the sequence of equalities applied at the third step of the algorithm is the sequence in Eqs. (A.5.1)–(A.5.5), we know that the sequences X that are obtained verify inequalities (A.1)–(A.3) (see Result 1). According to the fourth step of the algorithm, the sequence X obtained when the algorithm stops verifies inequality (A.4), and X is feasible. Considering step 5 of the algorithm and Result 2, the sequence X obtained when the algorithm stops corresponds to a sequence of idle windows that is not a subsequent sequence of a sequence of idle windows leading to a feasible solution. The result provided in the next section is based on this remark. Note that step 3 is executed at most Q times, where $Q = \sum_{i=1}^{m} q_i$ is the number of idle windows. Thus, the complexity of the algorithm is $O(Q(m + \log_2(Q)))$.

Remark

In step 5, if the idle period is included in the current one — that is, $[\alpha_{i,r_i+1}, \beta_{i,r_i+1}] \subset [\alpha_{i,r_i}, \beta_{i,r_i}]$ — it is not necessary to test a sequence that contains this period. We have to consider the next one — that is, $[\alpha_{i,r_i+2}, \beta_{i,r_i+2}]$ — assuming that this period is not included in $[\alpha_{i,r_i}, \beta_{i,r_i}]$.

A.2.2.4. Optimality

Result 3

The first feasible solution obtained by applying the algorithm is optimal.

Proof

As mentioned in the proof of Result 2, if S^* is a subsequent sequence of windows of S, then $x_i^* \geq x_i$, for $i = 1, \ldots, m$, where x_i^* is obtained starting from S^* while x_i is obtained starting from S. In particular, $x_{m+1}^* \geq x_{m+1}$, which completes the proof.

A.2.3. EXAMPLE

Consider the following example. A part should undergo three types of operation denoted 1–3. The first type of operation can be performed using three resources, the second one using two resources, and the third one using three resources. The part to be processed arrives at time 0. At this time, the idle periods are as follows:

Operation type 1
    Resource 1: [0, 2]; [4, 9]; [13, 15]; [25, +∞)
    Resource 2: [0, 4]; [8, 17]; [20, +∞)
    Resource 3: [4, 8]; [13, 16]; [20, 35]; [40, +∞)
Operation type 2
    Resource 1: [1, 9]; [11, 17]; [21, +∞)
    Resource 2: [2, 7]; [11, 13]; [22, +∞)
Operation type 3
    Resource 1: [8, 15]; [22, 33]; [35, +∞)
    Resource 2: [7, 11]; [22, 28]; [30, 35]; [40, +∞)
    Resource 3: [4, 15]; [18, 22]; [30, +∞)

A.2. THE LINEAR MANUFACTURING SYSTEM

For each type of operation, we order the periods as shown previously:

Operation type 1

$S_1 = \{[0, 4]; [0, 2]; [4, 9]; [4, 8]; [8, 17]; [13, 16]; [13, 15]; [20, +\infty);$
$[20, 35]; [25, +\infty); [40, +\infty)\}$

Operation type 2

$S_2 = \{[1, 9]; [2, 7]; [11, 17]; [11, 13]; [21, +\infty); [22, +\infty)\}$

Operation type 3

$S_3 = \{[4, 15]; [7, 11]; [8, 15]; [18, 22]; [22, 33]; [22, 28]; [30, +\infty);$
$[30, 35]; [35, +\infty); [40, +\infty)\}$

Assume that the time spent on the part should be as follows:

- To undergo operation 1: between 2 and 3 units of time
- To undergo operation 2: between 3 and 4 units of time
- To undergo operation 3: between 2 and 4 units of time

We develop the different steps of the algorithm (the first step is the order of the idle periods performed previously):

2. We set $r_1 = r_2 = r_3 = 1$
3. Sequence building
   $t_1 = 0$
   $t_2 = \max(1, 2 + 0) = 2$
   $t_3 = \max(4, 3 + 2) = 5$
   $t_4 = 2 + 5 = 7$
   $x_4 = 7$
   $x_3 = \max(5, 7 - 4) = 5$
   $x_2 = \max(2, 5 - 4) = 2$
   $x_1 = \max(0, 2 - 3) = 0$
4. Inequality checking
   $x_2 = 2 < \beta_{1,1} = 4$
   $x_3 = 5 < \beta_{2,1} = 9$
   $x_4 = 7 < \beta_{3,1} = 15$

As can be seen, none of the $i \in \{1, 2, 3\}$ satisfy $x_{i+1} > \beta_{i,r_i}$. Thus, the algorithm ends and the solution is as follows:

The part undergoes operation type 1 during period [0, 2].
The part undergoes operation type 2 during period [2, 5].
Finally, the part undergoes operation type 3 during period [5, 7].
$x_4 = 7$ is the minimal makespan for this part.

Only one iteration was necessary in this case. If, for instance, the first idle window of operation type 2 had been [1, 4] instead of [1, 9], we would have had $x_3 = 5 > \beta_{2,1} = 4$, and we would have had to restart the computation with the sequence [0, 4]; [2, 7] and [4, 15] of idle windows, and so on.

A.3. GENERALIZATION

A.3.1. NOTATIONS AND PROBLEM FORMULATION

Each operation of the process is defined by its minimal and maximal processing times. For instance, operation i is characterized by θ_i, the minimal processing time, and by $\theta_i + \delta_i$, the maximal processing time.

Each operation i is characterized by the set of its predecessors denoted by Γ_i^- and the set of its successors denoted by Γ_i^+. In this appendix, $j \in \Gamma_i^-$ means that j ends exactly when i starts. Similarly, $j \in \Gamma_i^+$ means that j starts as soon as i ends. The set of operations that begin at the same time as operation i is denoted by Φ_i^-, and the set of operations that end at the same time as i is denoted by Φ_i^+. In other words, if $i \in \Gamma_k^+$, then any $j \in \Gamma_k^+$ belongs to Φ_i^-. Similarly, if $i \in \Gamma_k^-$, then any $j \in \Gamma_k^-$ belongs to Φ_i^+.

Thus, a process P is characterized by the list $\{1, 2, \ldots, n\}$ of its operations, with each operation $i \in \{1, 2, \ldots, n\}$ characterized by six elements: $\{\theta_i, \delta_i, \Gamma_i^-, \Phi_i^-, \Phi_i^+, \Gamma_i^+\}$. We denote by b_i the starting time and by e_i the completion time of operation i, $i \in \{1, 2, \ldots, n\}$. At this point, the introduction of Φ_i^- and Φ_i^+ may appear to be somewhat useless since it is possible to use Γ_k^- and Γ_k^+ to define these variables. Nevertheless, these notations will considerably simplify the following explanations.

Note that if $i \in \Gamma_k^-$ and $j \in \Gamma_k^-$, then for any k' such that $i \in \Gamma_{k'}^-$, we also have $j \in \Gamma_{k'}^-$ since no delay is allowed between two consecutive operations. Similarly, if $i \in \Gamma_k^+$ and $j \in \Gamma_k^+$, then for any k' such that $i \in \Gamma_{k'}^+$, we also have $j \in \Gamma_{k'}^+$. Furthermore, since no operation begins at the same time that it finishes, $j \notin \Phi_i^-$ if $j \in \Gamma_i^+$, and $j \notin \Phi_i^+$ if $j \in \Gamma_i^-$.

The manufacturing system is composed of several resources, and the resources performing the same type of operation are identical. Since other products have been scheduled previously, the resources may be partially busy when a new product is introduced in the system. Since the system under consideration is managed under the no-wait constraint, semifinished products or components are not stored during the process. The only available flexibility in the system is given by the δ_i values that express the fact that the manufacturing time θ_i of operation i can be extended to $\theta_i + \delta_i$.

A.3. GENERALIZATION

This is in some way equivalent to the storage of semifinished products or components but makes the resource unavailable for other operations.

As in the linear case, we have a series of available windows denoted by $[\alpha_1^i, \beta_1^i], [\alpha_2^i, \beta_2^i], \ldots, [\alpha_{q_i}^i, \beta_{q_i}^i]$ for each operation i. These windows may be available on different resources. For each operation i, they are ordered in the increasing order of the α_k^i, $k = 1, 2, \ldots, q_i$. When two α_k^i are equal, the order is the increasing order of the β_k^i. There are no conflicts between the resources; in other words, the same resource cannot be used to perform different operations for the same product. The last window q_i, associated with each operation of type i, is such that $\beta_{q_i}^i = +\infty$.

A feasible solution to the problem associated with a set $[\alpha_{r_i}^i, \beta_{r_i}^i]$ of idle windows (where $i = 1, 2, \ldots, n$ and $r_i = 1, 2, \ldots, q_i$) is a set of operations whose starting and completion times are respectively b_i and e_i, $i = 1, 2, \ldots, n$. These times verify, for $i = 1, 2, \ldots, n$

$$\alpha_{r_i}^i \leq b_i \tag{A.11}$$

$$e_i \leq \beta_{r_i}^i \tag{A.12}$$

$$\theta_i \leq e_i - b_i \tag{A.13}$$

$$e_i - b_i \leq \theta_i + \delta_i \tag{A.14}$$

$$b_i = e_j, \forall j \in \Gamma_i^- \tag{A.15}$$

$$b_i = b_j, \forall j \in \phi_i^- \tag{A.16}$$

$$e_i = e_j, \forall j \in \phi_i^+ \tag{A.17}$$

$$e_i = b_j, \forall j \in \Gamma_i^+ \tag{A.18}$$

Note that inequalities (A.13) and (A.14) allow us to assign to operation i a processing time $w_i \in [\theta_i, \theta_i + \delta_i]$ such that $b_i + w_i = e_i$. A feasible solution is optimal if the makespan $\max_{i/\Gamma_i^+=\emptyset}(e_i)$ is minimal.

The goal is to schedule the operations required to manufacture a product whose process is P in order to minimize the production time (i.e., to complete the product as soon as possible). This scheduling has to be performed as soon as the product enters the system. Due to the intensity of the flow of products, products that have been previously scheduled cannot be rescheduled.

The scheduling problems with no-wait and blocking are two particular cases of the problem presented in this appendix. In the first case (no-wait problem) $\delta_i = 0$, whereas $\delta_i = +\infty$ in the blocking problem.

A.3.2. OPTIMAL SOLUTION FOR THE ACYCLIC PRODUCTION SYSTEM

A.3.2.1. Modeling and Test of Cyclicity

The system at hand can be represented by a precedence graph G, where

1. Each arc represents one operation type, and we call i the arc representing operation type i.
2. If $j \in \Gamma_i^-$, then the end of the arc representing j is the origin of the arc representing i. This means that operation of type i should start as soon as operation of type j ends. Similarly, if $j \in \Phi_i^-$, then the origin of j is the origin of i. In other words, operation types i and j start simultaneously. If $j \in \Phi_i^+$, then the end of j is the end of i (i.e., operation types i and j end simultaneously). If $j \in \Gamma_i^+$, then the origin of j is the end of i, or j starts as soon as i ends.

The precedence graph G is the set of arcs $\{1, 2, \ldots, n\}$, with each arc $i \in \{1, 2, \ldots, n\}$ being connected to the arcs belonging to $\{\Gamma_i^- \cup \Phi_i^- \cup \Phi_i^+ \cup \Gamma_i^+\}$. Thus, G can be defined as follows: $G = (\{\Gamma_i^-, \Phi_i^-, \Phi_i^+, \Gamma_i^+\}_{i=1,2,\ldots,n})$.

An undirected path of G is a sequence $i_1, i_2, \ldots, i_p$ of p arcs distinct from each other, such that either $i_{k+1} \in (\Gamma_{i_k}^- \cup \Phi_{i_k}^-)$ or $i_{k+1} \in (\Phi_{i_k}^+ \cup \Gamma_{i_k}^+)$ for $k = 1, 2, \ldots, p - 1$ and such that i_k and i_{k+2} are not connected to the same end of i_{k+1} for $k = 1, 2, \ldots, p - 2$. A cycle of G is an undirected path $i_1, i_2, \ldots, i_p$ (with $p \geq 2$) such that either $i_1 \in (\Gamma_{i_p}^- \cup \Phi_{i_p}^-)$ and $i_1 \in (\Phi_{i_2}^+ \cup \Gamma_{i_2}^+)$ or $i_1 \in (\Phi_{i_p}^+ \cup \Gamma_{i_p}^+)$ and $i_1 \in (\Gamma_{i_2}^- \cup \Phi_{i_2}^-)$. A cyclic precedence graph contains a cycle, whereas an acyclic precedence graph does not.

The goal of algorithm 2 presented later is to number the arcs of the precedence graph G such that the number associated with an arc i is greater than the numbers associated with the arcs connected with one end of i. Indeed, two arcs will have two different numbers.

Algorithm 2: Number the Arcs of an Acyclic Precedence Graph

1. Initialization
   1.1. $p = 1$. *p is a counter used to number the arcs.*
   1.2. For $i = 1, 2, \ldots, n$, set $m(i) = 0$. *When $m(i) = 0$, arcs are said to be unmarked. Thus, at the beginning of the algorithm, all the arcs are unmarked.*
2. While there exists i such that $m(i) = 0$ and at least one of the four following conditions is verified:
   i. $(\Gamma_i^- \cup \Phi_i^-) = \emptyset$: *The origin of arc i is not connected with another arc.*

A.3. GENERALIZATION

ii. $(\Phi_i^+ \cup \Gamma_i^+) = \emptyset$: *The end of arc i is not connected with another arc.*

iii. For any $j \in (\Gamma_i^- \cup \Phi_i^-)$, $m(j) \neq 0$: *All arcs that either end or start when arc i starts are marked.*

iv. For any $j \in (\Phi_i^+ \cup \Gamma_i^+)$, $m(j) \neq 0$: All arcs that either end or start when arc i ends are marked.

2.1. Set $m(i) = p$: *Arc i is marked.*

2.2. Set $p = p + 1$: *The counter is increased by one, which guarantees that the arcs will be marked following the increasing order of the positive integers.*

3. If some of the $m(i)$ are still at 0, then the graph is cyclic; otherwise, it is acyclic.

Remark

The algorithm starts with the initial numbers i assigned to the operations and generates an order $m(i)$, which is required by the approach presented in this appendix.

Result 4

If, at the end of algorithm 1, there exists $i \in \{1, 2, \ldots, n\}$ such that $m(i) = 0$, then the graph is cyclic. Otherwise, the graph is acyclic.

Proof

a. If, at the end of algorithm 2, there exists $i \in \{1, 2, \ldots, n\}$ such that $m(i) = 0$, then the graph is cyclic.

Assume that, at the end of algorithm 2, there exists $i \in \{1, 2, \ldots, n\}$ such that $m(i) = 0$. Arc i cannot be marked because there exists $j_1 \in (\Gamma_i^- \cup \Phi_i^-)$ and $k_1 \in (\Phi_i^+ \cup \Gamma_i^+)$ such that $m(j_1) = 0$, $m(k_1) = 0$, $j_1 \neq i$, and $k_1 \neq i$. Let us select j_1. Using the same reasoning, we find $j_2 \in (\Gamma_{j_1}^- \cup \Phi_{j_1}^-)$ such that $m(j_2) = 0$, $j_2 \neq j_1$, and the two arcs i and j_2 are not connected to the same extremity of j_1. In this way, we build an unlimited sequence $i = j_0, j_1, j_2, \ldots$ of arcs. Since the number of arcs in the graph is limited, this sequence will contain two arcs j_r and j_s such that $j_r = j_s$. Thus, the sequence of arcs $j_r, j_{r+1}, \ldots, j_{s-1}$ is a cycle.

b. If, at the end of algorithm 2, there does not exist $i \in \{1, 2, \ldots, n\}$ such that $m(i) = 0$, then the graph is acyclic.

Assume that the graph G contains a cycle $i_1, i_2, \ldots, i_p$ and that at the end of algorithm 1 there does not exist i such that $m(i) = 0$. Since $m(i_p) \neq 0$, we claim that either $m(i_1) < m(i_p)$ or $m(i_{p-1}) < m(i_p)$ to be allowed to mark i_p. If $m(i_1) < m(i_p)$, then $m(i_2) < m(i_1)$ and, step by step, we find that $m(i_p) < m(i_{p-1}) < \cdots < m(i_2) < m(i_1) < m(i_p)$. Similarly, if $m(i_{p-1}) < m(i_p)$, we obtain $m(i_p) < m(i_1) < \cdots < m(i_{p-2}) < m(i_{p-1}) < m(i_p)$. In both cases, we obtain $m(i_p) < m(i_p)$, which is impossible.

This completes the proof.

Result 4 establishes that algorithm 2 converges, even if the precedence graph is cyclic. According to the previous algorithm, if the graph is acyclic an arc is marked either if each operation j that begins or ends when i begins has been marked previously or if each operation j that begins or ends when i ends has been marked previously. As a consequence, if the graph is acyclic, each arc i verifies at least one of the following conditions:

1. Each operation j that begins or finishes when the operation i begins is such that $m(j) < m(i)$.
2. Each operation j that begins or finishes when the operation i ends is such that $m(j) < m(i)$.

Result 5
The complexity of algorithm 2 is in $O(n)$.

Proof
A node of the graph is one of the ends of an arc. The common end of two arcs connected to each other is considered a unique node. A way to work out algorithm 1 is to weight each node of the graph with the number of arcs that end or start at each node. An arc $i \in \{1, 2, \ldots, n\}$ can be marked if one of its ends is weighted at 1 since, in this case, all the arcs that end or start at this end are already marked, except the arc under consideration. When marking an arc i, we subtract 1 to the weights of each of its ends. We continue the process, selecting at each step one arc with one of its ends weighted at 1. At most, n steps are necessary to complete the process. Note that each time we subtract 1 to the weight of each of the ends of an arc, we test if one of these weights is equal to 1. If so, we store this arc in a particular list that contains the arcs ready for marking. Thus, the selection of an arc that can be marked is straightforward. Finally, algorithm 1 consists of (i) subtracting 1 to two weights at most n times and (ii) storing an arc in a special list at most n times. Thus, Result 2 holds.

A.3.2.2. Computing an Optimal Solution

In the remainder of this section, we assume that the precedence graph is acyclic and we use algorithm 2 to number the arcs. The numbers assigned to the operations are given in Fig. A.1. Then, we apply algorithm 3 in order to obtain an optimal solution (i.e., a solution, that minimizes the makespan).

A.3. GENERALIZATION

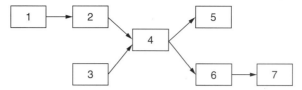

FIGURE A.1 A process composed of seven operations.

The idea behind algorithm 3 is quite simple and can be explained using Fig. A.1. Assume that the window to which each operation is assigned is known. In this process, there is an assembly–disassembly operation (operation 4). The others are processing operations. Assume that the window where operation i should be located is $[\alpha_{r_i}^i, \beta_{r_i}^i]$, for $i \in \{1, 2, \ldots, 7\}$.

We compute a_i, the lower bound of the starting time of operation i, and d_i, the lower bound of the completion time of operation i, for $i \in \{1, 2, \ldots, 7\}$. We obtain successively

Operation 1: $a_1 = \alpha_{r_1}^1$ and, since θ_1 is its minimal processing time, $d_1 = a_1 + \theta_1$.
Operation 2: Since operation 1 must be finished before beginning operation 2, $a_2 = \max(\alpha_{r_2}^2, d_1)$ and $d_2 = a_2 + \theta_2$.
Operation 3: $a_3 = \alpha_{r_3}^3$ and $d_3 = a_3 + \theta_3$.
Operation 4: $a_4 = \max(\alpha_{r_4}^4, d_2, d_3)$ and $d_4 = a_3 + \theta_3$.

For operations 1–4, we computed a_i, the lower bound of the starting times of the operations, first and based on these bounds we computed d_i, the lower bound of the completion times of the operations. The problem is quite different when considering operation 5 since a_5, the lower bound of the starting time of this operation, depends not only on d_4 but also on a_6, the lower bound of the starting time of operation 6, which is unknown at this stage of the computation. This is the reason why we reverse the computation process and compute d_5, the lower bound of the completion time of operation 5, first. We obtain

Operation 5: $d_5 = \alpha_{r_5}^5 + \theta_5$ and, since $\theta_5 + \delta_5$ is its maximal processing time, $a_5 = \max(\alpha_{r_5}^5, d_5 - \theta_5 - \delta_5)$.
Operation 6: Since operations 5 and 6 must start at the same time, $a_6 = \max(\alpha_{r_6}^6, d_4, a_5)$ and $d_6 = a_6 + \theta_6$.
Operation 7: $a_7 = \max(\alpha_{r_7}^7, d_6)$ and $d_7 = a_7 + \theta_7$.

We then apply a reverse process starting from d_7 to define a set of feasible starting times of the operations. This process leads successively to b_i, the minimal starting time of operation i, and to e_i, the minimal completion time of operation i, for $i \in \{1, 2, \ldots, 7\}$. Indeed, $a_i \leq b_i$ and $d_i \leq e_i$. Applying the reverse process, we obtain successively

Operation 7: $e_7 = d_7$ and $b_7 = a_7$.
Operation 6: Since operation 6 must end when operation 7 starts, $e_6 = \max(d_6, b_7)$, and since $\theta_6 + \delta_6$ is its maximal processing time, $b_6 = \max(a_6, e_6 - \theta_6 - \delta_6)$.
Operation 5: $b_5 = \max(a_5, b_6)$ and, since θ_5 is its minimal processing time, $e_5 = \max(d_5, b_5 + \theta_5)$.
Operation 4: $e_4 = \max(d_4, b_6)$ and $b_4 = \max(a_4, e_4 - \theta_4 - \delta_4)$.
Operation 3: $e_3 = \max(d_3, b_4)$ and $b_3 = \max(a_3, e_3 - \theta_3 - \delta_3)$.
Operation 2: $e_2 = \max(d_2, b_4)$ and $b_2 = \max(a_2, e_2 - \theta_2 - \delta_2)$.
Operation 1: $e_1 = \max(d_1, b_2)$ and $b_1 = \max(a_1, e_1 - \theta_1 - \delta_1)$.

If operation i cannot be completed before $\beta^i_{r_i}$ (i.e., if $e_i > \beta^i_{r_i}$), we replace window $[\alpha^i_{r_i}, \beta^i_{r_i}]$ by the next one—that is, $[\alpha^i_{r_i+1}, \beta^i_{r_i+1}]$—and we restart the process. We can speed up the computation by replacing the window $[\alpha^i_{r_i}, \beta^i_{r_i}]$ with the first window $[\alpha^i_{r_i+k_i}, \beta^i_{r_i+k_i}]$ (i.e., k_i is as low as possible and $k_i \geq 1$) such that the operation i can be completed before $\beta^i_{r_i+k_i}$. If all the operations can be completed in the windows to which they are assigned, then the b'_i values are the solution to the problem. This simple example illustrates the general algorithm proposed hereafter.

Algorithm 3: Acyclic Precedence

In the remainder of this section, i should be understood as $m(i)$ obtained by applying algorithm 2.

1. Initialization
   1.1. For $i \in \{1, 2, \ldots, n\}$, set $r_i = 1$
   r_i is the rank of the idle time window that is under consideration to perform operation i.
   1.2. For $i \in \{1, 2, \ldots, n\}$, set $f(i) = 0$
   $f(i)$ is the operation on which the computation of the schedule of operation i will be based.
2. Forward process: Computation of lower bounds for the starting times and the completion times of operations
   2.1. For $i = 1$ to n,
   2.1.1. Set $E_i = \{i+1, i+2, \ldots, n\}$
   E_i is the set of arcs that have not been consider yet.
   2.1.2. $a_i = \alpha^i_{r_i}$
   a_i is a lower bound of the starting time of operation i. Initially, a_i is set at the value of the lower limit of the window corresponding to operation i.

A.3. GENERALIZATION

2.1.3. $d_i = \alpha^i_{r_i} + \theta_i$
d_i is a lower bound of the completion time of operation i.
Initially, d_i is set at the value of the lower limit of the window corresponding to operation i increased by the lower value of the manufacturing time.

2.1.4. If $(\Gamma_i^- \cup \Phi_i^-) \cap E_i = \emptyset$, then
This condition means that, for each operation j that begins or finishes when operation i begins, a_i and d_i have been previously computed.

2.1.4.1. $a_i = \max(a_i, \max_{j \in \Gamma_i^-}(d_j) \max_{j \in \Phi_i^-}(a_j))$
The starting time a_i is the maximal value among (i) a lower bound of a_i, (ii) the greater completion time of the operations that precede i, and (iii) the greater starting time of the operations that are supposed to start at the same time as i.

2.1.4.2. $d_i = \max(d_i, a_i + \theta_i)$
The completion time d_i is the maximal value among (i) a lower bound of d_i and (ii) the sum of the starting time of operation i and the minimal processing time of i.

2.1.4.3. For $j \in (\Gamma_i^- \cup \Phi_i^-)$, do $f(j) = i$.
Any operation j that begins or finishes when operation i begins will be scheduled taking into account the starting time of operation i.

2.1.5. If $(\Phi_i^+ \cup \Gamma_i^+) \cap E_i = \emptyset$, then
This condition means that for each operation that begins or finishes when operation i finishes, a_i and d_i have been previously computed.

2.1.5.1. $d_i = \max(d_i, \max_{j \in \Phi_i^+}(d_j) \max_{j \in \Gamma_i^+}(a_j))$
The completion time d_i is the maximal value among (i) a lower bound of dd_i, (ii) the greater completion time of the operations that are supposed to end at the same time as i, and (iii) the greater starting time of the operations that succeed to i.

2.1.5.2. $a_i = \max(a_i, d_i - \theta_i - \delta_i)$
The starting time a_i is the maximal value among (i) a lower bound of a_i and (ii) the completion time of operation i minus the maximal processing time of i.

2.1.5.3. For $j \in (\Phi_i^+ \cup \Gamma_i^+)$, do $f(j) = i$
Any operation j that begins or finishes when operation i finishes will be scheduled taking into account the completion time of operation i.

3. Backward process
   Computation of b_i, the minimal starting time, and e_i, the minimal completion time of operation $i \in \{1, 2, \ldots, n\}$
   3.1. For $i = n$ down to 1,
      3.1.1. If $f(i) = 0$, then
      Operation i is scheduled before any operation that begins or finishes when operation i begins or finishes.
         3.1.1.1. $b_i = a_i$
         3.1.1.2. $e_i = d_i$
      3.1.2. If $f(i) \in \Gamma_i^-$, then
      Operation i is scheduled taking into account $e_{f(i)}$, the minimal completion time of operation $f(i)$ that has been previously scheduled and that ends when operation i begins.
         3.1.2.1. $b_i = \max(a_i, e_{f(i)})$
         3.1.2.2. $e_i = \max(d_i, b_i + \theta_i)$
      3.1.3. If $f(i) \in \Phi_i^-$, then
      Operation i is scheduled taking into account $b_{f(i)}$, the minimal starting time of operation $f(i)$ that has been previously scheduled and that begins when operation i begins.
         3.1.3.1. $b_i = \max(a_i, b_{f(i)})$
         3.1.3.2. $e_i = \max(d_i, b_i + \theta_i)$
      3.1.4. If $f(i) \in \Phi_i^+$, then
      Operation i is scheduled taking into account $e_{f(i)}$, the minimal completion time of operation $f(i)$ that has been previously scheduled and that ends when operation i ends.
         3.1.4.1. $e_i = \max(d_i, e_{f(i)})$
         3.1.4.2. $b_i = \max(a_i, e_i - \theta_i - \delta_i)$
      3.1.5. If $f(i) \in \Gamma_i^+$, then
      Operation i is scheduled taking into account $b_{f(i)}$, the minimal starting time of operation $f(i)$ that has been previously scheduled and that begins when operation i ends.
         3.1.5.1. $e_i = \max(d_i, b_{f(i)})$
         3.1.5.2. $b_i = \max(a_i, e_i - \theta_i - \delta_i)$
4. Test
   4.1. For $i = 1$ to n,
      4.1.1. While $e_i > \beta_{r_i}^i$, set $r_i = r_i + 1$
   Each operation i must be completed before $\beta_{r_i}^i$. Therefore, we have to try the next time window.
   4.2. If none of the r_i has been modified, then
      4.2.1. $\max_{i / \Gamma_i^+ = \emptyset}(e_i)$ is the minimal makespan

4.2.2. b_i is the earliest starting time of the operations
4.3. Else
4.3.1. Return to step 2.

Remark

In algorithm 3, it is assumed that for each operation i, the corresponding windows are ordered in the increasing order of $\alpha_{r_i}^i$ and, in case of equality, in the increasing order of $\beta_{r_i}^i$. Under this assumption, it can be seen that

- In the forward process, a_i and d_i are computed once for each operation i and used once to compute $a_{f(i)}$ and $d_{f(i)}$; therefore, the complexity is in $O(n)$.
- In the reverse process, we compute a_i and d_i once for each operation i, which leads to complexity in $O(n)$.
- The previous computation is performed at most q times, where $q = \sum_{i=1}^{n} q_i$ is the total number of idle windows.

Finally, the complexity of the whole algorithm is in $O(qn)$.

Results 6 and 7 presented next guarantee that the solution obtained by applying this algorithm is optimal.

Result 6

Algorithm 3 leads to a feasible solution.

Proof

This proof can be found in Chauvet *et al.* (1998).

Result 7

In the first feasible solution obtained by applying algorithm 2, all the completion times and the starting times of the operations are minimal. As a consequence, this solution is optimal.

Proof

This proof can be found in Chauvet *et al.* (1998).

A.3.3. ILLUSTRATIVE EXAMPLE

The process represented in Fig. A.1 is composed of seven operations. Each operation is defined by θ_i, its minimal processing time, and $\theta_i + \delta_i$, its maximal processing time (Table A.1). Components cannot be stored in front of the resources. In this example, operations (except operation 1, which is a chemical treatment) can be extended as much as necessary, but

TABLE A.1 Processing Times

Processing times	Operation						
i	1	2	3	4	5	6	$n=7$
θ_i	1	2	3	3	2	1	2
δ_i	1	$+\infty$	$+\infty$	$+\infty$	$+\infty$	$+\infty$	$+\infty$

TABLE A.2 Available Time Windows (Idle Periods)

Time window	Operation						
i	1	2	3	4	5	6	$n=7$
q_i	1	3	3	3	3	2	4
$[\alpha_1^i, \beta_1^i]$	$[1, +\infty)$	$[0, 3]$	$[0, 2]$	$[0, 2]$	$[0, 5]$	$[3, 8]$	$[0, 3]$
$[\alpha_2^i, \beta_2^i]$		$[4, 11]$	$[6, 13]$	$[3, 7]$	$[7, 17]$	$[10, +\infty)$	$[5, 7]$
$[\alpha_3^i, \beta_3^i]$		$[14, +\infty)$	$[15, +\infty)$	$[10, +\infty)$	$[20, +\infty)$		$[10, 15]$
							$[17, +\infty)$

in this case machines are unavailable for other operations until the current operation ends.

Each operation is assigned to a given machine. Since some operations have been previously scheduled, the seven operations must be performed in one of the idle periods $[\alpha_{r_i}^i, \beta_{r_i}^i]$ of the resources (Table A.2).

By applying algorithm 3, we obtain the optimal solution in which the starting times of operations are minimal. This optimal solution is represented in Fig. A.2.

A.4. WORK-IN-PROCESS REGULATION

In this section, we consider an assembly system that includes $K + 1$ production stages. Note that the approach could be applied to a more general type of production system. Each production stage $k (k = 0, 1, \ldots, K)$ is composed of p^k pools of machines, with each pool $p = 1, 2, \ldots, p^k$ containing m_p^k machines. s_p^k storage entities are ready to store one unit of product in front of the pth pool of the kth production stage. If the pool is composed of assembly resources, each storage entity can store all the components required to assemble one unit of the product or semifinished product resulting from the assembly operation. The production stages are numbered from 0 (final production stage) to K. If several components belonging to the same product are manufactured in the same production stage, different pools of machines manufacture them.

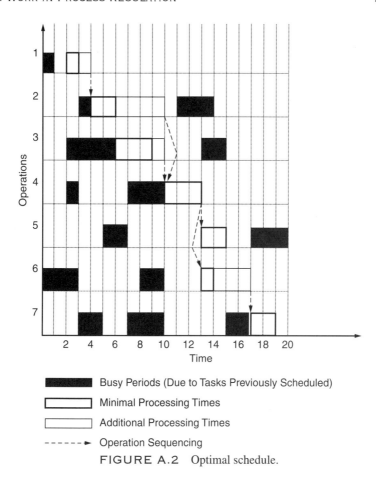

FIGURE A.2 Optimal schedule.

The assembly process can be performed on the assembly system represented in Fig. A.3. The pools P_p^k associated with the operations are the pools of machines required to perform the operations.

In such an assembly system, the processing time associated with each operation is fixed (i.e., $\delta_i = 0$ for any manufacturing or assembly operation i). Each storage entity is considered a resource. The minimal processing time for such a resource is equal to zero (i.e., $\theta_i = 0$ if i is a "storage operation"), and the maximal processing time is the same for each storage operation (i.e., $\delta_i = \delta$ for any storage operation i).

The parameter δ is used to control the work-in-process. It is expected that the lower the δ, the lower the production cycle, but the lower the utilization ratio of the machines. The goal of the following numerical examples is to test this assumption.

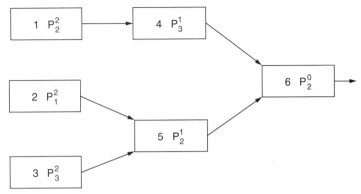

FIGURE A.3 An assembly process and the pools.

TABLE A.3 Number of Pools per Stage

Stage	5	4	3	2	1	0
Number of pools	6	2	5	9	5	2

To perform these tests, we generate assembly processes, at random and in sequence, and we schedule the operations as soon as the assembly processes are generated. This is done for different values of δ. For each value δ, we compute the utilization ratio of the bottleneck pool of machines (URBP). This ratio reflects the productivity of the system. We also compute the mean value of the production cycle time (MPCT) as well as the mean utilization ratio of the storage facilities (URST). There are six operation stages in the example ($K = 5$). Table A.3 gives the number of pools for each stage.

Each pool is made with three identical machines, and four storage identities are associated with each pool. For each manufacturing or assembly operation, the processing time is generated at random between 0 and 10. The assembly process is also generated at random, starting from level 0. The number of predecessors of each operation is bounded, as is the total number of operations in a manufacturing process. As a consequence, some of the manufacturing processes have less than six operation stages.

We generate 1000 manufacturing processes, launched in production in the order they were generated. For each value of δ, ($\delta = 0, 1, \ldots, 10$), we provide the MPCT (this value indicates, on average, the tendency of the cycle times), URST, and URBP.

The results are given in Table A.4. The conclusion is straightforward: The greater the δ, the greater the productivity (since the value of the URBP increases with δ), but the greater the level of the work-in-process (WIP) (since the URST increases with δ) and the greater the cycle time.

TABLE A.4 Control of the Work-in-Process

δ	MPCT	URST	URBP
0	22.03	0.000	0.722
1	24.67	0.050	0.776
2	27.34	0.104	0.818
3	30.23	0.162	0.858
4	33.03	0.219	0.884
5	35.96	0.276	0.906
6	38.80	0.331	0.923
7	41.51	0.379	0.931
8	44.08	0.422	0.944
9	46.37	0.460	0.950
10	48.45	0.490	0.953

A.5. CONCLUSION

The algorithms presented in this appendix are real-time scheduling algorithms that allow one to schedule new assembly while minimizing makespan. As a consequence, the first idle periods used are the ones that are the closest to the current time. This guarantees good use of the resources. Furthermore, by controlling the flexibility of the system (i.e., the δ_i values), it is possible to reduce the WIP at the expense of the use of resources. In particular, as shown in the example presented in Section A.4, it is possible to adjust the WIP according to the required productivity. Note that a set of storage locations is assigned to each pool of machines. This implies that the state of each one of these pools should be known at each time. Finally, this appendix clearly links the WIP level with the productivity and the production cycle.

The most interesting aspect is explained in Section A.4, in which it is shown how this real-time algorithm can be used to reduce the WIP at the expense of the productivity of the system. We are at the heart of supply chains.

REFERENCES

Baker, K., *Introduction of Sequencing and Scheduling.* Wiley, New York, 1974.

Callahan, J. R., "The nothing hot delay problems in the production of steel," PhD thesis, Department of Industrial Engineering, University of Toronto, 1971.

Chauvet, F., Levner, E., Meyzin, L. K., and Proth, J. M.,"On-line part scheduling in a surface treatment system," INRIA research reports No. 3318. INRIA, Le Chesnay, France, 1997.

Chauvet, F., Proth, J.-M., and Wardi, Y., "On-line scheduling with WIP regulation," Rensselaer's International Conference on Agile, Intelligent and Computer Integrated Manfacturing, Troy, NY, October 7–9, 1998.

Chu, C., Proth, J.-M., and Wang, L., "Improving job-shops schedules through critical pairwise exchanges," *Int. J. Production Res.*, **36**(3), 638–694, 1998.

Hall, N. G., and Sriskandarajah, C., "A survey of machine scheduling problems with blocking and no-wait in process," *Operations Res.*, **44**, 510–525, 1996.

McCormick, S. T., Pinedo, M. L., Shenkers, S., and Wolf, B., "Sequeucing in an assembly line with blocking to minimize cycle time," *Operations Res.*, **37**, 925–935, 1989.

Rachamadugu, R., and Stecke, K., "Classification and review of FMS scheduling procedures," *Production Planning Control* **5**(1), 2–20, 1994.

Song, W., Zabinsky, Z. B., and Storch, R. L., "An algorithm for scheduling a chemical processing tank line," *Production Planning Control* **4**(4), 323–332, 1993.

Index

Accounting methods, 81
Acer, 9
AIX, 129
AMD, 97
Apple Computer, 129
Application software, 130–132
Availability, 89–90

Barnes and Noble, 150
Benetton, 10–12
Best Buy, 55
British Steel (BS), 10
Buy activity, strategic level
   defined, 17
   goals and examples of, 17–18
   information processing, 47, 49–50
   mathematical model, 44–47
   sensitivity analysis, 51–54
Buy activity, supply chain design and
   components identified and analyzed, 25, 27
   sharing of benefits and losses, 32–33
Buy activity, tactical level defined, 65
   production cycle, minimizing time on, 68

Capacity requirement planning (CRP), 136–137
Cash flow, 89
Caterpillar, 55
Central processing units (CPUs), 97, 126
Chips, computer, 126
Christopher, M., 3
Chrysler Corp., 14
Cincinnati Milcron, 55
Circuit City, 55
Commonality, 109–110
Compaq Computer, 9, 55, 97
Competition, product, 117
Computer-aided design (CAD), 112, 130, 136
Computers, applications
   enterprise resource planning, 137–144
   Internet business applications, 150–152
   legacy, 135–137
   supply chain planning, 145–149
Computers, development of, 125
   business need and, 132–134
   hardware, 126–127
   human resources and, 134–135

Computers, development of *(continued)*
  networks, 127–128
  software, 128–132
Conceptual design, 106–108
Contingency tables, 91, 92
Copacino, William C., 3
Corporate culture, product development and, 103
Correspondence analysis, 91
Cost analysis, 81–84, 87–89
  deferment design and, 110
  product development and, 110
Cost for quality (CQ), 77–79
Customer(s)
  demographic and psychological profiles of, 76
  expectations, 90–91
  product development and role of, 104, 105–106
  service, improving, 85–86
  service, measuring, 91, 93
Customer relationship management (CRM), 150–151
Customer satisfaction, 66
  evaluating, 89–93

Daimler-Chrysler, 14, 98
Decision making
  global consequences, 62–65
  supply chain design and analysis of, 28–29
Decision-making systems (DMSs), 1, 2
Deere, John, 55
Deferment design, cost analysis of, 110
Dell Computer Corp., 9–10, 97
Demand visibility, improving, 75–76
Design for assembly (DFA), 112
Design for manufacture (DFM), 111–112
Design for manufacture and assembly, 112
Design for "X" (DFX), 111–112
Digital, 8, 128
Dominant partners, 54–58

E-commerce, 150
Electronic data interchange (EDI), 137
Enterprise resource planning (ERP), 137–144

Financial performance, evaluation of, 87–89
Fine, Charles H., 3–4
Ford Motor Co., 14
Forecasting, 75–76
Free-form fabrication, 113

Gattorna, John, 3
General Motors, 14

Hardware, computer, 126–127
Hewlett-Packard (HP), 9, 129
Human resources, affects of computers on, 134–135

IBM, 8, 9, 128, 129, 150
Information flow, 8
Information processing
  constraints, 47–48
  department organization and, 48–50
  in supply chain environment, 50–51
Information sharing, 31–32
Intel, 9, 97
International Standard for Organization (ISO), 80
Internet, 132, 133–134, 150–152
i2 Technologies, 146

Japan, steel industry in, 12–14
JAVA, 129
Joy, Bill, 126
Just in time (JIT), 73

Kaizen, 80
K-Mart, 55

Legacy computer applications, 135–137
Linear manufacturing system, 161–168
Local area networks (LANs), 127

Index

Make activity, strategic level
  defined, 17
  goals and examples of, 18
  information processing, 47, 48, 49
  mathematical model, 43–44
  sensitivity analysis, 51–54
Make activity, supply chain design and, 22
  components identified and analyzed, 25, 26, 27
  external events analyzed, 30
  sharing of benefits and losses, 33
Make activity, tactical level
  defined, 65
  production cycle, minimizing time on, 69
Manufacturing requirements planning (MRP-II), 137
Manugistics, 146
Marketing, product
  strategies, 102, 114–115
  test, 114
Master production scheduling (MPS), 136
Master-slave basis, 9, 10, 73
Materials requirement planning (MRP), 136
Mathematical model
  basic assumptions, 35
  buy activity, 44–47
  example, 177–181
  information processing, 47–51
  linear manufacturing system, 161–168
  make activity, 43–44
  move activity, 38–40
  notations and problem formulation, 168–169
  sell activity, 36–38
  sensitivity analysis, 51–54
  solution for acyclic production system, 170–177
  store activity, 41–42
Merrill Lynch, 150
Microsoft, 8, 9, 129
Million instructions per (MIPS), 126

Module, use of term, 20
Moore, Gordon, 126
Move activity, strategic level
  defined, 17
  goals and examples of, 18–19
  information processing, 47, 48, 50
  mathematical model, 38–40
  sensitivity analysis, 51–54
Move activity, supply chain design and
  components identified and analyzed, 25, 27
  sharing of benefits and losses, 33
Move activity, tactical level
  defined, 66
  production cycle, minimizing time on, 69
MS-DOS, 129

Networks, computer, 127–128
Nokia, 150

Operating systems, computer, 128, 129
Operational performance, evaluation of, 89–93
Oracle, 139
OS/2, 129

PeopleSoft, 139
Performance evaluation, 86–93
Personal digital assistants (PDAs), 126
Physical systems (PSs), 1, 2
Pipeline visibility, improving, 73–75
Point of differentiation (PD), 108–109
Product
  design and development, 103–114
  identification, 100–103
  introduction, 114–116
  phaseout, 118–120
  sustenance, 116–117
Product development
  architecture, 108–110
  conceptual design, 106–108
  cost analysis, 110
  customer requirements and, 105–106

Product development *(continued)*
    description of, 103–114
    detailed design, 111–112
    logistical problems, 97–98
    marketing strategies, 102
    outsourcing, 98
    prototyping and testing, 112–114
    system design for, 110–111
Production life cycle
    defined, 99–100
    minimizing time on, 68–73, 104
    stages in, 99–120
Production system, design of, 110–111
Programming languages, 128–129
Prototyping, product, 112–114

Quality
    assurance, 80
    improving, 77–81
    mastery, 77–79
    total, 80–81
Quality function deployment (QFD), 106

Random access memory (RAM), 126
Rapid prototyping, 113
Reengineering, 80–81
Renault, 80–81
Research and development (R&D)
    financial evaluation and, 88
    subcontracting, 88
Return on investment (ROI), 66–67, 88–89
R/3 ERP system, 140

SAP AG, 139, 140
Schary, Philip B., 4
Scheduling, real-time, 69
Sears, 150
Sell activity, strategic level
    defined, 17
    goals and examples of, 19–20
    information processing, 47, 48, 49, 50
    mathematical model, 36–38
    sensitivity analysis, 51–54

Sell activity, supply chain design and, 22
    components identified and analyzed, 25, 27
    decision making consequences analyzed, 28
    sharing of benefits and losses, 34
Sell activity, tactical level
    defined, 66
    production cycle, minimizing time on, 69
Sensitivity analysis, 51–54
Services, improving, 85–86
Sharing process, 7–8, 15
    of benefits and losses, 30–35
    of information, 31–32
Skjott-Larsen, Tage, 4
Software, computer
    application, 130–132
    system, 128–130
Solaris, 129
Sony, 150
Star system, 129
Store activity, strategic level
    defined, 17
    goals and examples of, 19
    information processing, 47, 48, 50
    mathematical model, 41–42
    sensitivity analysis, 51–54
Store activity, supply chain design and, 22
    components identified and analyzed, 25, 27
    sharing of benefits and losses, 33
Store activity, tactical level
    defined, 66
    production cycle, minimizing time on, 69
Strategic Alignment Model, 3
Strategic level
    activities in, 17–20
    dominant partners, 54–58
    mathematical model, 35–54
    supply chain design, 21–35
Sun Microsystems, 129
Supply chain planning (SCP), 145–149

Supply chains
    comparison of examples, 15
    computer applications for, 145–149
    definition of, 7–8
    examples of, 9–16
    goals of, 155, 157
    research on, 2–5
Supply chains, design of, 21
    basics of, 22–24
    components identified and analyzed, 24–27
    decision making consequences analyzed, 28–29
    external events analyzed, 29–30
    sharing of benefits and losses, 30–35
    steps for, 24
System software, 128–130

Tactical level
    activities in, 65–66
    decision making and global consequences, 62–65
    performance evaluation, 86–93
Tactical level, objectives, 66–67
    converting orders into cash, minimizing time on, 68–73
    costs, reducing, 81–84
    demand visibility, improving, 75–76
    pipeline visibility, improving, 73–75
    quality, improving, 77–81
    services, improving, 85–86
    work-in-process, minimizing total, 73
Technology
    See also Computers, development of
    disruptive, 103
Tesco, 55, 150
Testing, product, 112–114
Toshiba, 97, 150
Total quality management (TQM), 80–81

UNIX, 128–129

Vendor Managed Inventory, 76

Wal-Mart, 55
Wide area networks (WANs), 127
Windows 95, 129
Windows NT, 129
Work-in-process (WIP), 73, 159, 178–181

Xerox, 129